# Community Connections For Science Education

## VOLUME I

## BUILDING SUCCESSFUL PARTNERSHIPS

*By William C. Robertson*

**Art and Design**
Linda Olliver, Director
**NSTA Web**
Tim Weber, Webmaster
**Periodicals Publishing**
Shelley Carey, Director
**Printing and Production**
Catherine Lorrain-Hale, Director
**Publications Operations**
Erin Miller, Manager
**sciLINKS**
Tyson Brown, Manager

**National Science Teachers Association**
Gerald F. Wheeler, Executive Director
David Beacom, Publisher

NSTA Press, NSTA Journals, and the NSTA Web site deliver high-quality resources for science educators.

Shirley Watt Ireton, Director
Beth Daniels, Managing Editor
Judy Cusick, Associate Editor
Linda Olliver, Cover Design

---

*Community Connections for Science Education: Building Successful Partnerships*
NSTA Stock Number: PB160X1
ISBN 0-87355-191-5
Library of Congress Control Number: 200190648
Printed in the USA by Kirby Lithographic Company, Inc.
Printed on recycled paper

Copyright © 2001 by the National Science Teachers Association.
The mission of the National Science Teachers Assocation is to promote excellence and innovation in science teaching and learning for all.

Permission is granted in advance for reproduction for purpose of classroom or workshop instruction. To request permission for other uses, send specific requests to:

NSTA Press
1840 Wilson Boulevard
Arlington, Virginia 22201-3000
www.nsta.org

# Table of Contents

**About the Author** .................................................................................................. v

*National Science Education Standards* **Matrix** ............................... vi

**An NSTA Position Statement for Informal Science Education** . viii

**Chapter 1 Introduction** ...................................................................................... 1

**Chapter 2 What's out there?**
    Informal Science Education Opportunities ................................................. 7
    Formal Science Education Opportunities ................................................. 14

**Chapter 3 Getting started—deciding what you want**
    Questions Informal Sites Should Answer .................................................. 19
    Questions Formal Educators Should Answer ........................................... 26

**Chapter 4 Making it happen**
    Initial Contact ............................................................................................... 29
    Educational Materials .................................................................................. 30
    Teacher and Staff Training .......................................................................... 33
    Administrative Concerns ............................................................................. 35

**Chapter 5 The visit**
    Recommendations for Formal Educators ................................................. 39
    Recommendations for Informal Sites ........................................................ 43

**Chapter 6 Maintaining the relationship**
    Key Ingredients ............................................................................................ 47

**Web Resources** .................................................................................................... 55

**Appendix A**
    About Parks as Resources for Knowledge in Science (PARKS) ............... 59

**Appendix B**
    Directory of PARKS Participants .............................................................. 61

# About the Author

William C. Robertson is an assistant professor of physics turned curriculum developer who has a master's degree in physics and a Ph.D. in science education. He has numerous publications on issues ranging from conceptual understanding in physics to how to bring constructivism into the classroom. Bill has developed K–12 science curricula, teacher materials, and award-winning science kits for Biological Sciences Curriculum Study, The United States Space Foundation, The Wild Goose Company, and Edmark. He is currently a freelance science education writer, reviewer of science materials, and teacher of online math and physics at the university level.

## Acknowledgments

This book and its partner, *Community Connections for Science Education: History and Theory You Can Use*, are part of a project funded by the National Park Foundation with support from ExxonMobil. *Building Successful Partnerships* was reviewed and enhanced by the expertise of Christie Anastasia and Patti Reilly of the National Park Service. These books were planned and developed by Julia Washburn of the National Park Foundation and Shirley Watt Ireton of the National Science Teachers Association. At NSTA, the project editor was Beth Daniels; Linda Olliver designed the book layout and cover; and Jack Parker, Nguyet Tran, and Catherine Lorrain-Hale handled production.

We would like to thank all of the formal and informal educators who contributed their time and valuable information to the making of this book. Their input made this effort possible.

# *National Science Education Standards* Matrix

Science Teaching, Professional Development, Assessment, Program, and System Standards

| Standard | A | B | C |
|---|---|---|---|
| **Science Teaching** | Plan inquiry-based science program for their students (p. 30). | Guide and facilitate learning (p. 32). | Engage in ongoing assessment of their teaching and of student learning (p. 37). |
| **Professional Development** | Requires learning essential science content through the perspectives and methods of inquiry (p. 59). | Requires integrating knowledge of science, learning, pedagogy, and students; it also requires applying that knowledge to science teaching (p. 62). | Requires building understanding and ability for lifelong learning (p. 68). |
| **Assessment** | Assessments must be consistent with the decisions they are designed to inform (p. 78). | Achievement and opportunity to learn science must be assessed (p. 79). | The technical quality of the data collected is well matched to the decisions and actions taken on the basis of their interpretation (p. 83). |
| **Education Program** | All elements of the K-12 science program must be consistent with the other *National Science Education Standards* and with one another and developed within and across grade levels to meet a clearly stated set of goals (p. 210). | The program of study in science for all students should be developmentally appropriate, interesting, and relevant to students' lives; emphasize student understanding through inquiry; and be connected with other school subjects (p. 212). | The science program should be coordinated with the mathematics program to enhance student use and understanding of mathematics in the study of science and to improve student understanding of mathematics (p. 214). |
| **Education System** | Policies that influence the practice of science education must be congruent with the program, teaching, professional development, assessment, and content standards while allowing for adaptation to local circumstances (p. 230). | Policies that influence science education should be coordinated within and across agencies, institutions, and organizations (p. 231). | Policies need to be sustained over sufficient time to provide the continuity necessary to bring about the changes required by the *Standards* (p. 231). |

| D | E | F | G |
|---|---|---|---|
| Design and manage learning environments that provide students with the time, space, and resources needed for learning science (p. 43). | Develop communities of science learners that reflect the intellectual rigor of scientific inquiry and the attitudes and social values conducive to science learning (pp. 45–46). | Actively participate in the ongoing planning and development of the school science program (p. 51). | |
| Professional development programs for teachers of science must be coherent and integrated (p. 70). | | | |
| Assessment practices must be fair (p. 85). | The inferences made from assessments about student achievement and opportunity to learn must be sound (p. 86). | | |
| The K–12 science program must give students access to appropriate and sufficient resources, including quality teachers, time, materials, and equipment, adequate and safe space, and the community (p. 218). | All students in the K–12 science program must have equitable access to opportunities to achieve the *National Science Education Standards* (p. 221). | Schools must work as communities that encourage, support, and sustain teachers as they implement an effective science program (p. 222). | |
| Policies must be supported with resources (p. 232). | Science education policies must be equitable (p. 232). | All policy instruments must be reviewed for possible unintended effects on the classroom practice of science education (p. 233). | Responsible individuals must take the opportunity afforded by the standards-based reform movement to achieve the new vision of science education portrayed in the *Standards* (p. 233). |

National Research Council. 1996. *National science education standards.* Washington, D.C.: National Academy Press.

# An NSTA Position Statement on Informal Science Education

## Preamble

NSTA recognizes and encourages the development of sustained links between the informal institutions and schools. Informal science education generally refers to programs and experiences developed outside the classroom by institutions and organizations that include:

- Children's and natural history museums; science-technology centers; planetaria; zoos and aquaria; botanical gardens and arboreta; parks; nature centers and environmental education centers; and scientific research laboratories
- Media, involving print, film, broadcast, and electronic forms
- Community-based organizations and projects, including youth organizations and community outreach services

A growing body of research documents the power of informal learning experiences to spark curiosity and engage interest in the sciences during school years and throughout a lifetime. Informal science education institutions have a long history of providing staff development for teachers and enrichment experiences for students and the public. Informal science education accommodates different learning styles and effectively serves the complete spectrum of learners: gifted, challenged, nontraditional, and second-language learners.

## Declaration:

NSTA strongly supports and advocates informal science education because we share a common mission and vision articulated by the National Science Education Standards:

- Informal science education complements, supplements, deepens, and enhances classroom science studies. It increases the amount of time participants can be engaged in a project or topic. It can be the proving ground for curriculum materials.
- The impact of informal experiences extends to the affective, cognitive, and social realms by presenting the opportunity for mentors, professionals, and citizens to share time, friendship, effort, creativity, and expertise with youngsters and adult learners.
- Informal science education allows for different learning styles and multiple intelligences and offers supplementary alternatives to science study for non-traditional and second language learners. It offers unique opportunities through field experiences, field studies, overnight experiences, and special programs.
- Informal science learning experiences offer teachers a powerful means to enhance both professional and personal development in science content knowledge and accessibility to unique resources.
- Informal science education institutions, through their exhibits and programs, provide an effective means for parents and other care-providers to share moments of intellectual curiosity and time with their children.
- Informal science institutions give teachers and students direct access to scientists and other career role models in the sciences, as well as to opportunities for authentic science study.
- Informal science educators bring an emphasis on creativity and enrichment strategies to their teaching through the need to attract their noncompulsory audiences.
- NSTA advocates that local corporations, foundations, and institutions fund and support informal science education in their communities.
- Informal science education is often the only means for continuing science learning in the general public beyond the school years.

*—Adopted by the NSTA Board of Directors in January 1998*

## References

Beane, D. B. 1990. Say yes to a youngster's future: a model for home, school, and community partnership. *Journal of Negro Education* 59(3): 360–374.

Bergstrom, J. 1984. *School's out—Now what?* Berkeley, Calif.: Ten Speed Press.

Billingsley, A., and C. Caldwell. 1991. The church, the family, and the school in the African American community. *Journal of Negro Education* 60(3): 427–440.

Bloom, B. 1981. *All our children learning: A primer for parents, teachers, and other educators.* New York: McGraw-Hill.

Crane, V., H. Nicholson, M. Chen, et al. 1994. *Informal science learning: What research says about television, science museums, and community-based projects.* Dedham, Mass.: Research Communications, Ltd.

Dierking, L., and J. H. Falk. 1994. Family behavior and learning in informal science settings: A review of the research. *Science Education* 78(1): 57–72.

Druger, M. 1988. *Science for the fun of it: A guide to informal science education.* Arlington, Va.: National Science Teachers Association.

Kyle, W.C. 1984. Influence of school and home factors on learning. In *Research within reach; A research-guided response to the concerns of educators.* Edited by D. Holdzkom and P. B. Lutz. Washington, D.C.: National Science Teachers Association, 123–141.

Miller, S. R. and K. Pittman. 1987. *Opportunities for prevention: Building after-school and summer programs for young adolescents.* Washington, D.C.: Children's Defense Fund.

National Eisenhower Program for Mathematics and Science, U.S. Dept. of Education. 1982. *Science museums and school change.* Philadelphia: The Franklin Institute Science Museum.

National Science Foundation. 1997. Collaborators in reform. In *Foundations: A monograph published by the division of elementary, secondary, and informal education.* Arlington, Va.: Directorate for Education and Human Resources, 67–76.

St. John, M. 1996. *An invisible infrastructure: Institutions of informal science education.* Washington, D.C.: Association of Science-Technology Centers, Inc.

## Main Authors of Informal Science Position Statement

**Andrea Anderson**—*Educational Consultant*
**Marvin Druger**—*Past President, NSTA*
**Chuck James**—*Director, Carnegie Institution's Office of Science Education*
**Phyllis Katz**—*Director, Hands On Science Outreach*
**Jennifer Ernisse**—*Manager of Programs, McWane Center, Adventures in Science*
**Members of the NSTA Informal Science Advisory Board**

Chapter 1

# Introduction

This book is a guide for building partnerships between formal and informal science educators, so before we start, let's define what those terms mean. Formal science educators include teachers and administrators who are involved in the "formal" education of kids. In other words, people who toil away in schools of all kinds—public, private, and home. Informal science educators include the people and the facilities associated with museums, parks, zoos, aquariums, after-school enrichment programs, and even amusement parks.

The *National Science Education Standards*, first published in 1996, clearly outline the need to expand the borders of the traditional science classroom. As stated in the *Standards,* "The classroom is a limited environment. The school science program must extend beyond the wall of the school to the resources of the community....Many communities have access to science centers and museums, as well as to the science communities in higher education, national laboratories, and industry; these can contribute greatly to the understanding of science and encourage students to further their interests outside of school" (45). The National Science Teachers Association (NSTA) fully supports this position and realizes that such community resources can be of most benefit to students when strong, lasting partnerships are formed between formal and informal science educators.

This book and its partner, *Community Connections for Science Education: History and Theory You Can Use*, are part of a project funded by the National Park Foundation, with support from ExxonMobil, called Parks As Resources for Knowledge in Science (PARKS). PARKS is a partnership of the National Park Service, the National Park Foundation, the National Science Teachers Association, and Ohio State University. The PARKS goals are as follows:

- Integrate the *National Science Education Standards* into National Park curriculum-based education programs.
- Create a model framework for integrating the *Standards* into other park programs (e.g., history or social studies programs).
- Promote the National Parks as learning laboratories and provide opportunities for students and teachers to use National Park resources to achieve the *National Science Education Standards*.

## Introduction

- Enhance the quality of science education for students nationwide.
- Increase knowledge of, and stewardship for, National Park resources, the National Park System, and associated resources.
- Establish Parks As Classrooms® (National Park Service education program) models that can be replicated at other national parks, while fostering the incorporation of the *National Science Education Standards* into existing and future National Park Service education programs.

Now, if you are an educator at a site that is not associated with the National Park Service, or are an educator not within easy reach of a national park, rest assured that this book addresses all kinds of informal science education sites. Some of the contributors to this book are associated with PARKS, but many are not. PARKS has an established model for facilitating formal-informal partnerships, so we will describe what they do in some detail.

Beginning in 1998, PARKS chose 32 parks to receive $25,000 grants over a two-year period and to send a teacher/ranger team to one of four *National Science Education Standards* workshops presented jointly by the National Park Service, NSTA, and Ohio State University. The goals of these workshops fit into three major categories: (1) modeling and teaching the *National Science Education Standards*, including helping sites plan and implement student activities appropriate for each site; (2) providing and supporting, through follow-up visits, professional development for both formal and informal educators; and (3) helping sites and educators develop plans for evaluating their educational programs.

Among the many recommendations in the *Standards* are teaching science through inquiry, focusing on understanding rather than memorization, selecting appropriate curricula from various sources rather than relying on prefabricated programs, increasing teacher understanding, and assessing student learning in appropriate ways. The *Standards* also contain specific recommendations on science content. As a formal or an informal science educator, you might be asking yourself why it's necessary to tie what you're doing to the *Standards*. After all, there is no such thing as a national science curriculum. First, the kind of science recommended in the *Standards* is just good stuff! Other than that, though, the *Standards* closely correlate with what most states and districts require in their local standards. That's no accident, by the way. The developers of state and local standards used the *National Science Education Standards* as their starting point, tailoring the work to reflect state and local priorities. Bottom line: The *National Science Education Standards* reflect what schools want happening in science classrooms across the country.

Professional development is an important part of the PARKS program and is essential if you wish to establish a formal-informal partnership in science education.

# Chapter 1

Professional development workshops were designed for *both* kinds of educators. In the PARKS workshops, the organizers realized that the development process had to extend beyond the workshop itself, and so implemented a series of follow-up visits to the sites. The message for both sites and formal educators is that successful partnerships require more than arranging one-time interactions. We'll provide lots of examples of what that means in a later chapter.

> Before the workshop I was not familiar with the *National Science Education Standards*. From the workshop I gained a basic understanding of these and learned the importance of linking the *Standards* with the program. I have been able to do this in the development stages. This also allowed me to come back to the park and develop an understanding of how my own state standards fit in…
>
> —PARKS workshop participant

The final focus of the workshops was evaluation. Sites need to evaluate the effectiveness of their programs, and teachers need to evaluate student learning associated with a site visit. Developing appropriate evaluation tools is no easy task, and training people to do so requires time and effort. While some participants in the PARKS workshops felt a more theoretical approach to evaluation was adequate, many needed more specific instruction on developing evaluation tools and programs. One thing we hope to do in this book is provide some specific recommendations for evaluation of programs and student learning.

## Has this ever happened to you?

You might wonder why a book like this is necessary. After all, teachers plan field trips, informal sites hang around waiting for the kids, everyone learns something, and the buses leave. What could go wrong? To discover what could go wrong, we're going to paint two pictures of a field trip—one from the perspective of an eager classroom teacher, and one from the perspective of an eager volunteer at a regional park. Those two eager people are…you.

### A field trip nightmare

You're a first-year middle school teacher. Your progressive science class has just begun a brief excursion into cultural anthropology. Desperate to provide a rewarding experience for your class, you check out a directory of local museums for a field trip. The Natural History Museum is nearby, and they're offering a special educational exhibit—Hunting Tools Since the Dawn of Time. What luck! You call the museum and schedule a field trip for the following week.

# Introduction

Of course it's a hassle dealing with permission slips, arranging the bus transportation, but you hastily arrange things and are ready for the big day. Your class arrives at the museum and enters the lobby. Seeing that your guide has yet to meet you, you take charge of the situation. "Okay, everyone, if you have to go, do it now. There are the restrooms." Several kids scurry off but quickly return with news that the restrooms are out of order. Okay, you'll deal with that later. Ah, there's the guide.

A stern-faced man approaches you and tells you and the kids to follow him to the lecture hall. Everyone dutifully obeys, and then the guide begins in his best monotone. "Now before you learn anything about our exhibits, there are a few rules here. First, no talking. Giggling and carrying on is no way to behave in a museum. Second, I expect you to take notes on what I say. We use tests at the end of the day to determine whether or not your class has learned anything and can be invited back at a later date. Now before the tour, I'd like to tell you a little about the very important research that has been conducted here at the museum. Now if you will show the first slide, please. Yes, this shows the large amount of dirt we removed from the site before…" Despite your best intentions, you begin to nod off.

An hour later you awaken to silence. The guide is ready to hand you over to a colleague for the walking tour of the museum. While the first guide was a droner, this one talks so fast you can't keep up. After a quick admonition to "look, but don't touch," she speeds you and the class on a whirlwind tour of the exhibits. You get the feeling that if anyone interrupted her train of thought, she would have to start the whole spiel over again. Fearing that to be the case, you and the class remain mute.

After an hour of looking but not touching, the guide announces that you will head to the classroom for a hands-on activity. Well, finally. After a boring lecture and a light-speed tour, the kids are going to interact with something and maybe, just maybe, learn something from this trip. "Today is your lucky day, kids, because you're going to make your very own hunting sling!" The class groans. This is the exact same activity you did with them in preparation for the field trip. Why hadn't these people returned your follow-up calls so you would have known what to expect? Why did you rush into this field trip? The class has lunch, fails the museum's end-of-day test, and heads for the bus. Gee, what an enlightening day!

## And now the shoe is on the other foot...

You are a new staff member at Horned Owl Regional Park. Attendance at the park has steadily declined since that outbreak of a rare owl disease a few years ago, and management wants to recapture its audience through an outreach program with the local schools. Given your previous teaching experience, you are the logical choice to get this program off the ground.

You resolve to form a positive partnership with schools. You start by contacting local principals and setting up meetings to discuss the educational programs the park has to offer. On the first day of meetings, you arrive at Sandlot Elementary School for your scheduled appointment with the principal. She arrives 30 minutes later. You explain your programs—the hands-on activities, active student involvement, and programs designed to help the teachers and students meet their science standards. You also explain the requirements of the school—how many students you can handle at one time, what is expected of the teachers, etc. Halfway through this explanation, the principal stands up and says, "This sounds really great! We'll be sending teachers your way very soon. I'm so glad we had this meeting!" You leave all the materials teachers are supposed to use to prepare the kids for field trips and walk out the door with just a slight spring in your step.

One month later, you are expecting the first group of kids from Sandlot Elementary. As you wait in the lobby of the welcome center, you see the bus round the corner and enter Horned Owl Regional Park. Then you see another bus, and another. Now wait a minute. Why does it take three buses to cart around 30 kids and the required number of chaperones? Ah, here's why. Instead of following your guidelines of no more than 30 students, Sandlot has sent the entire fourth grade, consisting of 137 students. They sent three chaperones. One chaperone is 15 years old, a high school recruit. Another is a parent with a small child and a hyperactive beagle without a leash. Finally, a teacher emerges, and you engage him as to how everyone is going to survive the day. You quickly discover that this is a substitute teacher who has no knowledge of the park. He has no idea of the activities the students were supposed to do before coming.

Oh well, no need to panic. May as well make the best of it and head out to the first viewing site so the kids can begin investigating. As you begin the hike, you notice very few students prepared for hiking; sandals are the norm, and nobody seems to have a jacket for the rain that's on the horizon. Great. When you do reach the first site, chaos reigns. The kids begin running, jumping, and yelling all over the place. You catch one kid trying to set fire to one of the park benches!

# Introduction

> You manage to keep the kids from burning up the park, and you actually survive the morning. Fed up, you inform the teacher that, after lunch, the rest of the field trip is canceled. Your best move of the day! Whew! Thank goodness that's over. Calmly, you walk back to the welcome center and vow to revise your tactics. But what's this? Ah, four students who didn't make it back from lunch in time to get on the buses. Lovely.

## It doesn't have to be like that

Believe it or not, just about everything contained in those descriptions has happened to the people and sites that have contributed to this book. Of course, not all of those things happened on a single field trip, but there's a lot more truth than fiction in those stories. Things can go desperately wrong on both sides of the relationship if you don't plan ahead and know what you're getting into. And of course, that's one of the reasons we've written this book. We want to do our best to help you avoid the pitfalls others have taken, and we also want to share the positive things that can result when formal and informal educators form lasting partnerships.

In the next chapter, we'll discuss the kinds of informal science education opportunities that might be available in your area. We'll also profile science educators who have successfully worked with informal sites.

# Chapter 2

# What's Out There?

**B**ecause you are reading this book, we are going to make the bold assumption that you're interested in creating, improving, or solidifying a formal-informal partnership in science education. In this chapter, you will discover the opportunities that exist nationwide and in your locale for creating new partnerships or adding to your partnership repertoire. As with all the chapters in this book, we will address the issue both from the perspective of the formal educator and from the perspective of the informal educator.

## Informal Science Education Opportunities

If you live in or near a relatively large metropolitan area, there are numerous fairly obvious informal education sites available—science museums, natural history museums, planetariums, zoos, aquariums, and parks. But what if you live in a small town or are looking for new and unique sites to visit with your students? Well, there are endless opportunities in either case. Below is an incomplete list of possibilities.

- Visit the local water treatment plant for an enlightening, and mildly disgusting, peek into the science behind recycling your water supply.
- Contact the division of wildlife or the state game and fish commission to inquire about possible involvement of your students in animal tagging programs, fish stocking and breeding, or environmental cleanup.
- Any number of businesses in your area should be happy to have your students learn the science behind what they do. Try a recycling facility, food manufacturer, or computer chip manufacturer. In rural areas, a trip to a farm or ranch, accompanied by a qualified individual such as someone from the local department of agriculture or a cooperative university extension, could lead to an understanding of the science of crop rotation, soil conservation, or animal husbandry.
- For older students, amusement parks provide an entertaining look into the principles of physics. Just type "amusement park physics" into any Internet search engine, and you'll have more information than you can handle.

## What's Out There?

> Students tour the industrial site at Avery Island to learn how biology, chemistry, and physics are used in the production and bottling of Tabasco.

- Many areas have after-school science enrichment programs. Although most of these are for-profit companies that require student fees; some, such as ASPIRA (described in detail below), provide their own funding.
- Educational foundations whose main purposes are to promote understanding and awareness of science sites through education. Examples are the Chesapeake Bay Foundation, the Living Classrooms Foundation, and the United States Space Foundation.
- Local radio and television stations. The students can explore science topics, from the functioning of broadcast equipment to meteorology to the clever way they make it look like the weatherperson has a map directly behind him or her.

Of course, this is only a partial list. For more ideas—or to investigate some of these sources further—check some of the web sites listed at the back of this book. Now for a more detailed look at the kinds of things offered by informal science education sites, we profile below the sites that have contributed time, energy, and advice to help create this book.

**Students learn better through the practical, hands-on activities available in the informal education sites.**

At this point, we should mention one thing about the sites we profile in this book. They range in size from the small to the large, but don't include super-large sites that you might find in large metropolitan areas. Such informal science education sites typically have education departments with quite a large staff and many, many projects that involve science education at all levels. Unless you're dealing with a site in its first year or two of operation, chances are that a very large site has already established many partnerships and connections with local school districts as well as companies and educational organizations nationwide.

The fact that these large sites most likely already have long-standing relationships with formal educators doesn't mean you can't take advantage of the opportunities they offer. Your best bet is to call the site and get in touch with the education department. Explain that you want to involve your students in science education experiences outside the classroom. After being directed to the proper person in the department, you should be in business. The right person might direct you to someone in your school district who is in charge of an existing partnership, or he or she might be able to get you going.

## The ASPIRA Association

Formed in 1961, ASPIRA is a nationwide association whose goal is to develop the educational and leadership capacity of Hispanic youth. The organization takes its name from the Spanish verb *aspirar*, "to aspire." With a national office in Washington, D.C., and associate offices in New York, New Jersey, Pennsylvania, Illinois, Florida, Connecticut, and Puerto Rico, ASPIRA provides various educational, enrichment, and support services to Latino youth and their families in schools, community centers, and ASPIRA clubs. Associate offices either manage their own alternative schools or develop partnerships with schools, community organizations, businesses, community leaders, and parents. One of ASPIRA's programs is the Math and Science (MAS) Academy. Through this program, ASPIRA offers after-school enrichment activities during the school year and summer for middle school students. They have a series of weekly hands-on math and science sessions; career orientation; visits to scientific institutions; role modeling and interaction with minority scientists; and parent involvement in a family festival. The MAS Academy programs use the *National Science Education Standards* as a guide. ASPIRA supports its programs with the MAS Institute, which is a three-day training session for teachers that helps them implement the programs for students and parents and provides them with curriculum materials to help them meet the national standards in their classrooms.

> Students need interesting and challenging out-of-school experiences to enrich studies, pique curiosity, and provide motivation.

The MAS Academy recruits teachers and students from other ASPIRA programs, from neighboring schools, and through word of mouth. All that is required to participate in the program is an interest in math and science. You don't often find an easier entrance requirement! ASPIRA's website is *www.aspira.org*. There, you can find locale-specific information on all of their programs. Cost-free after-school enrichment, coupled with teacher training and parent involvement, is a valuable resource.

## Louisiana Public Broadcasting

Like most Public Broadcasting System (PBS) affiliates, Louisiana Public Broadcasting (LPB) offers links to the entire range of PBS productions and a full slate of instructional television programming for teachers, and others, to use free of charge. Unlike the other informal sites featured in this book, however, LPB does not seek to attract students to its site. LPB forms partnerships by helping teachers use its technical expertise in the areas of satellite-delivered courses, compressed video, streaming video, and broadcast. In addition, LPB collaborates with the Louisiana State Department of Education, Louisiana State University, the National Science Foundation, and the like, to apply for grants and implement projects that further science education in the state. Examples of the results of those collaborations include:

## What's Out There?

- SATELLITE SIX, a federally funded grant that delivered graduate-level content coursework for uncertified science and math teachers to help them towards certification.
- Dr. Dad's PH3, a 12-part series on CD-ROM designed to encourage girls to look at science as fun, interesting, and exciting.
- PBS Scienceline, a web-based resource for teachers that helps them understand and implement inquiry-based science teaching and other aspects of the *National Science Education Standards*.
- Enviro-Tacklebox™, a series of middle school video modules on environmental education that is based on the NSTA publication, *Decisions Based on Science*.

As an explicit part of their mission, LPB looks for grant opportunities and partners to help them meet their goals. You can find out more about their programs and contact them via their website at *www.lpb.org*.

> Most teachers just don't have the experience to write and submit grant applications. If an informal site can help with that process, then that's a significant help.

### The National Aquarium in Baltimore

This is one of the major public aquariums in North America. It is owned by the City of Baltimore and operated by a private, not-for-profit corporation, National Aquarium in Baltimore, Inc. There are two major buildings: the Main Aquarium and the Marine Mammal Pavilion. The Main Aquarium has four levels of galleries with exhibits ranging in size from 100 gallons (380 liters) to 10,000 gallons (37,800 liters), circling an open space with a 275,000-gallon (1,040,000-liter) pool full of rays, small sharks, and big bony fish. The building is topped by a 5,000 square foot (465 square meter) South American Rainforest and an Amazon River Forest exhibit, which show the diverse group of animals that live in the water among the trees during seasonal flooding. Other exhibits include the Western Atlantic Coral Reef (335,000 gallons/1,270,000 liters) and the Open Ocean/Sharks (225,000 gallons/852,000 liters). The Marine Mammal Pavilion houses a dolphin show stadium that presents dolphins with a conservation perspective. It also has museum-style exhibits on marine mammals and their adaptations, a children's discovery touch pool that allows hands-on experiences with marine invertebrates under the guidance of a trained staff member, and two formal classrooms for hands-on programs. During the school year, the Main Aquarium is the site of school programs that are keyed to the Maryland State Department of Education Science Content Standards and the Science Outcomes and Indicators, as well as the American Association for the Advancement of Science (AAAS) Benchmarks for Science Literacy. Schools that do not schedule a regular education program receive a package of pre-, during-, and post-visit materials keyed to the Maryland State Standards, Outcomes, and Benchmarks, allowing teachers to use the aquarium more effectively.

Roughly 200,000 students, teachers, and chaperones visit the aquarium each year, with half of those coming from the state of Maryland. Most Maryland students attend free due to a grant from the Maryland State Department of Education. The students covered by this grant include public, private, and home-schooled children. The education staff at the aquarium works with teachers, schools, school districts, and state science supervisors. They have an Adopt-a-School program that provides intensive teacher in-service, outreach, and in-house programming to particular Baltimore schools (two at a time) for several years. Grants from foundations such as the Howard Hughes Medical Institute provide teacher in-service to the Baltimore City Public Schools, and aquarium staff have served on city school textbook adoption committees. Through grants from the National Science Foundation, the education staff has developed and disseminated an aquatic science curriculum, *Living in Water*, both regionally and nationally.

The National Aquarium in Baltimore makes a significant effort to reach schools and teachers through mailings and a presence at regional science meetings, but you can contact them directly via their Website at *www.aqua.org*.

## *Pocono Environmental Education Center*

> The Standards for Professional Development require that teachers learn essential content through the perspectives and methods of inquiry; that teachers integrate knowledge of science, learning, pedagogy, and students, and that the knowledge of this be applied to science teaching; that understanding and ability for lifelong learning is built; and that the programs for teachers be coherent and integrated.
>
> —*National Science Education Standards* (NRC 1996, 55–72)

Located in the Delaware Water Gap National Park (DEWA), the Pocono Environmental Education Center (PEEC) advances environmental awareness, knowledge, and skills through education. PEEC's classroom is 100 thousand hectares of national park and proximal public land on the Pocono Plateau in rural Pike County, Pennsylvania. PEEC formed a partnership with the National Park Service in the early 1970s and continues as a member of the PARKS program. The programs that PEEC provides include the following:

- Hands-on natural science opportunities in a national park
- Professional development for teachers and instructors
- Residential environmental education program for schools, Scout groups, and families
- *Ecotones*, a quarterly science education newsletter for educators

# What's Out There?

- Acting as a local, national, and international clearinghouse for environmental education and programs
- A Toyota USA Foundation-sponsored Summer Leadership Institute for 20 of the nation's outstanding science teachers

PEEC uses an inquiry-based approach to science, in which students choose and conduct their own environmental research projects. The students use an extensive set of process skills, first to gather data, and then to analyze it with the guidance of a mentor.

With more than half a million participants in over 25 years, PEEC continues to offer year-round institutes and workshops, ranging in topics from bird watching to photography to nature study to environmental education. Their programs support not only the *National Science Education Standards,* but also state science standards for Pennsylvania, New York, and New Jersey. You can contact them at *www.peec.org* or by calling 570-828-2319.

## Santa Monica Mountains National Recreation Area

Located in the Los Angeles and Ventura Counties, Santa Monica Mountains National Recreation Area (SAMO) provides a place for teachers to take students where they can engage in meaningful field science. SAMO has more than 50 miles of coastline and 580 miles of trails for hiking, mountain biking, and horseback riding. With its close proximity to Los Angeles and its many educational institutions, the area provides a good place for students and scientists to learn firsthand about urban-wildland interactions; ecosystems; plants and animals; and the effect of urban areas on water, air quality, and biodiversity. All of the recreation area's education programs are curriculum based and were developed with teachers. The goals of the education programs are:

- To introduce and motivate students to learn about the major themes of the Santa Monica Mountains National Recreation Area.
- To introduce students to the National Park Service mission of preservation and protection of natural and cultural resources.
- To meet the academic needs of students and educators in the Los Angeles and Ventura County school systems.
- To develop public support for the management of the National Park System and SAMO.
- To introduce students to the outdoors.
- To introduce students to career opportunities.

All of SAMO's science education programs are tied to the *National Science Education Standards* and the California Benchmarks in science, math, and other relevant subjects. Their science education programs include the following:

- The Chumash: A Changing People, Changing Land. This is a hands-on program for third- and fourth-grade students that explores biodiversity and the limits of natural resources through the question, "Could the Chumash people (early inhabitants of the area) live on the land today in the way they did for thousands of years?"
- Parks As Laboratories: Studies of the Land, Water, and Air. This program for sixth- through eighth-grade students targets primarily urban students who have little contact with natural settings such as SAMO. The students conduct "health exams" of park sites through water, soil, and air-quality measurements.
- National Park Labs: Studies of Wildland Fire Ecology. This is a program for ninth- and tenth-grade students that integrates scientific concepts across disciplines to learn about fire ecology and resource management. The students conduct hands-on field experiments, record data, and analyze the data.

For more than 10 years, SAMO has worked closely in partnership with the Los Angeles Unified School District. School district and park-based funding sustain park programs. Teacher workshops and training have been conducted with support from California Lutheran University and the University of California at Los Angeles. SAMO advertises its programs through mailings and other traditional outlets, but it specifically targets students in economically underserved areas. The SAMO website is *www.nps.gov/samo/*.

## *Shenandoah National Park*

Shenandoah National Park sits along the Appalachian Trail and Skyline Drive in western Virginia. Its diverse ecosystem is a reservoir for the area's native plants and animals, and its watersheds preserve the area's water quality and scenic values. Known publicly for its outdoor recreational opportunities, the park used a grant from the Parks As Classrooms initiative to develop an educational program starting in 1991. The park has programs for students in grades K through 12 that are tied to the Virginia Standards of Learning and the *National Science Education Standards*. Shenandoah produces a semiannual education program newsletter that reaches more than 500 local educators, providing program updates, ideas and resources, and evaluation results. The park also provides programs and workshops for university and adult audiences, in cooperation with James Madison University College of Education and College of Integrated Science and Technology.

Shenandoah has a work group entirely dedicated to education programs, which works with teachers and other partners to develop appropriate and relevant curriculum-based programs. The park holds four to six teacher workshops each year to provide participants with guidance and instruction on implementing the park's programs.

Shenandoah National Park has established a particularly strong partnership with McGaheysville Elementary School. All 500 of the school's students attend a park

## What's Out There?

program during the year, and the school hosts an open house that showcases what the students have learned with the park throughout the year.

You can contact the park through their website at *www.nps.gov/shen/* or by calling 540-999-3500.

## Formal Science Education Opportunities

To someone in charge of education at an informal education site, it might seem funny to think that certain kinds of formal educators represent "opportunities." After all, there are lots and lots of teachers, schools, and students out there to attract to your site. Well, for starters, let's just take a look at why you want to get *any* formal educators to your site.

The most obvious reason is that reaching out to formal educators is an easy way to further the science education of your local community, however large you consider that community to be. If the mission of your site isn't all about furthering science education, then you wouldn't be reading this book!

A second reason to reach out to formal educators is that it will probably increase attendance at your site from the general public. Students who have a good experience are likely to bring their families back for a second or third visit. Also, when word gets out that you're doing good things for schools, the community as a whole is more likely to patronize your valuable establishment.

A third reason is to promote any educational items you sell to keep your site afloat. These can be curriculum materials intended for the classroom or retail items you have in your gift shop or on your website. Students who have done a cool hands-on activity as part of a field trip are likely to pick up a science kit that contains that activity and others. Remember, if you think it crass to introduce commercialism into this venture, consider that if your site doesn't meet the bottom line, you won't be furthering *anyone's* understanding of science.

Now, in the first paragraph of this section, we mentioned that *certain* formal educators represented opportunities. Just as it is possible for a teacher to make a bad choice of a site for a field trip, it is also possible for a site to make a bad choice in forming a partnership with a teacher. The word *partnership* is key here. We're talking about a long-standing relationship—be it with a teacher, school, or district—that results in visits that are predictable, stable, and reliable. In other words, no unpleasant surprises. Your goal is to educate the community, not see how stressed out you can make you and your staff. We'll have specific recommendations for how to choose formal educators in a later chapter. For now, we'll provide profiles of formal educators who have formed successful partnerships with informal sites and whom we'll hear from later.

## Andrea Bowden, Supervisor, Office of Science, Mathematics and Health Education, Baltimore City Public Schools

Andrea R. Bowden is Supervisor K–12 of the Office of Science, Mathematics and Health Education for the Baltimore City Public School System (BCPSS). During her 32 years with BCPSS, she has served in several capacities as a central office administrator and as a science teacher in five high schools. She coordinates curriculum production, assessment, professional development, instructional support, special programs, and environmental programs. Additionally, she manages major federal grant initiatives such as National Science Foundation Teacher Enhancement in Mathematics, Safe and Drug Free Schools, and AIDS-HIV Education; and state-funded grant initiatives such as Eisenhower Professional Development, Service Learning, and Environmental Education. She has directed the NSF-funded summer SANDALS program at the College of Notre Dame of Maryland to prepare students for careers in science, mathematics and technology. Dr. Bowden has also taught graduate courses for Johns Hopkins University.

Dr. Bowden was the first Maryland recipient of the Presidential Award for Excellence in Science Teaching in 1983. Among her other awards are Outstanding Science Supervisor in Maryland and the Bene Merenti (for good works) Medal from the College of Notre Dame of Maryland. She has served in leadership capacities in a number of professional organizations at the state and national level, including president of both the Maryland Association of Science Teachers and Biology Teachers, District III Director of NSTA and chair of the National Science Teachers Regional Convention in Baltimore in November 2000. Dr. Bowden has been a member of many boards, including those of the National Aquarium and the Maryland Science Center. Recently the Chesapeake Bay Foundation named her the Outstanding Environmental Educator of the Year 2000.

## David W. Burchfield, Ed.D., Principal, McGaheysville Elementary School

David Burchfield has served as a principal of McGaheysville Elementary for seven years. Prior to becoming an administrator, he served as a first- and second-grade teacher, and administered an early childhood grant in Albemarle County, Virginia. He earned his doctorate in administration and supervision from the Curry School at the University of Virginia. David has served as an educational consultant and led workshops dealing with early childhood and elementary instructional practices, and for the past four years has been an instructor in Ohio at Muskingum College's "Educating Children Summer Training Institute." He has published articles and chapters on developmentally appropriate practices, and his classroom was featured on a NAEYC/NCREL video, "Developmentally Appropriate First Grade: A Community of Learners." He lives with his wife Bonnie and three daughters in Massanutten, Virginia, and can be reached at *dburchfield@rockingham.k12.va.us*.

# What's Out There?

### *Sheila Hodges, Science Teacher, Delaware Valley High School*

Sheila Hodges is a science teacher at Delaware Valley High School in Pike County in Northeastern Pennsylvania. The area is considered rural with a major urban influence due to its proximity to New York City. Sheila teaches biology and environmental science to students at the high school level, and is also the advisor of the Delaware Valley High School Environmental Club. Sheila was previously employed as an environmental specialist for a private engineering firm and as an archeologist. Hodges has both a BS and an MS in biology. She was voted the 1997 Conservation Educator of the Year and was the recipient of the 1999 Northeast Pennsylvania Environmental Partnership Award. Hodges and the Environmental Club received this award for their outstanding accomplishments and teamwork in achieving excellence in environmental protection through partnerships with the school, the community, and local businesses.

Sheila is also a Penn State Master Gardener and, as such, has volunteered over 180 hours of community work in the past year. DVHS Environmental Club members work with other Master Gardeners to clean up and restore various estates in the Delaware Water Gap National Recreation Area. Hodges has received her sixth consecutive grant in the past five years. The monies received are being used to purchase water quality equipment to aid in the research of local streams and lakes. Another grant was used to obtain a state of the art weather system monitored by local TV stations.

### *Henry Ortiz, Environmental Field Science Education Specialist, Los Angeles Unified School District*

Henry Ortiz works as one of several science advisors for the Los Angeles Unified School District. He coordinates most of the K–12 environmental education programs in that district. These programs include: Field Science Training for Teachers, The GLOBE program (Henry is the Los Angeles GLOBE program franchise coordinator), National Parks Studies In Wildland Fire Ecology, The Parks As Laboratories Program, Trout in the Classroom, The Temescal Canyon Field Science Program, The Yosemite and Eastern Sierra Teacher Institutes, and the Fort MacArthur Marine Science Program. Henry has a Biological Sciences background and he has taught at the high school and middle school levels. He is a science trainer for the GLOBE program and has trained teachers in Spanish and English in national and international trainings. He is also a trainer for the University of Hawaii's Fluid Earth/ Living Ocean oceanography and marine biology programs and for the UCLA Leadership In Marine Science (LIMS) and Science Standards with Integrated Marine Science (SSWIMS) programs. Henry serves on several committees, including the California Environmental Education Advisory Committee. One of his main goals is to ensure that every student in his school district is provided with an opportunity to participate in an environmental education program at one point throughout his or her academic journey.

# Chapter 2

### *Faimon Roberts, Assistant Director for Science, Louisiana State Systemic Initiatives Program*

Faimon Roberts taught middle school science and mathematics for 22 years before becoming the Assistant Director for Science with the Louisiana Systemic Initiatives Program (LaSIP). He has been the National Science Teachers Association District VII Director, Past-President of the National Middle Level Science Teachers Association, and the current treasurer and a Past-President of the Louisiana Science Teachers Association and president of the Louisiana Association of Science Leaders. Mr. Roberts directs the science professional development and leadership programs for LaSIP. During 1992–2000, he administered 123 science professional development projects (for $15.6 million) that have provided long-term professional development to over 4,700 teachers. He is also the lead person in the Developing Educational Excellence and Proficiencies (DEEP) in Science effort to provide science leaders to assist schools and districts in their efforts to increase student achievement.

## What's Out There?

Chapter 3

# Getting Started—Deciding What You Want

**B**efore you get yourself involved in a one-day or 10-year partnership, it might be a good idea to decide exactly what you want out of that partnership. Although we're not talking about marriage, there are a lot of similarities! If the formal and informal educators are looking for different kinds of relationships, trouble is on the horizon. And if either party goes into the relationship expecting the other to change, at least one party is in for a letdown. Just ask any radio talk show psychologist. In this chapter we pose numerous questions, then provide answers and suggestions given by the people currently involved in successful partnerships.

## Questions Informal Sites Should Answer

**1. What is your purpose in reaching out to formal educators? Given that purpose, who exactly is your audience, and how does that audience affect your approach?**

> In my experience, one of the most important aspects is to have a mutually agreeable goal or purpose—a relationship that has benefits for both partners. Bad relationships can be created if it benefits only one side or the other—this can lead to hard feelings. Understand how your mission and goals match with the partner's goals. Be clear about the partner's expectations. Don't force your mission and goals on the partner. On the other hand, don't assume you can meet every demand; determine what you can do based on staffing, budget, and site limitations and stick with it. Stress quality over quantity!
>
> — *Tim Taglauer, Shenandoah National Park*

The quotes in this section are typical of what we received from our informal educators. All agreed that it is a waste of time to determine your purpose and actions isolated from your intended audience. The overwhelming message is to find out what the schools want rather than try to shove your agenda down their collective throats. If

## Getting Started—Deciding What You Want

schools in your area want programs that tie to their district science standards, then a successful partnership should include that as a component. If they can only enter into a partnership that offers in-service credit for their teachers, you won't have much luck if that's not included in your program. If, however, they want you to baby-sit their students for a day, tell them to take a hike. Of course, the goals of your partnership must work both ways, and benefit both parties. This can't happen without communication.

> Communication and flexibility are the key elements. Find out what teachers need and provide it. Open, honest communication leads to fewer problems. Never assume; ask for clarification and find out what the stakeholders expect.
>
> —Al Staropoli, ASPIRA

> The partnership requires that we be free in our communication—both ways. We try to work within the context of what the Baltimore City Public Schools wants and needs. All of our programs take into account their curricula. We try to ask folks what they want rather than tell them what they need....Go meet teachers on their own ground. Work in school classrooms, go to teacher meetings, and get to know their needs.
>
> —Andrea Bowden, Baltimore City Public Schools (BCPS)

Assuming you have met with teachers and administrators in your area to determine your general focus for the partnership, you still have to determine your audience. For example, ASPIRA's primary goal is to further the educational and leadership opportunities of Hispanic students. In this task, they realize that the programs will fail without parental support. So the target audience of ASPIRA is not only students, teachers, and administrators, but also parents. They developed a specific program (the Family MAS festival) just for this purpose.

### Target Audiences

Whatever we offer has to be seen as something that adds to, rather than duplicates, what the teachers do in their classroom. *Al Staropoli, ASPIRA*

Do not assume what is needed by school systems, schools, and teachers—establish a dialogue and plan together. Show how the field experience will enhance student performance and help on tests, rather than being time "missed" from instruction. *Andrea Bowden, BCPS*

We have developed relationships through both teachers and administrators, but the greatest successes have been realized when both teachers and administrators value the relationship. First, deal with whoever has the

> most interest, but then be sure to get "buy in" from those "in charge."
> *Tim Taglauer, Shenandoah National Park*
>
> If you find the teaching leaders (not school administrators) and work with them, they will bring other teachers with them.... It is best to work with classroom teachers to ensure that they support participating in the program and that their students are prepared.... Before developing a program, however, it is crucial that you get the support of the school administrator.
> *Barbara Applebaum, Santa Monica Mountains National Recreation Area (SAMO)*
>
> Enhancing teachers' knowledge of environment is an important way to help improve student performance. The teacher is the key to student success. *John Padolino, PEEC*

Once again, our contributors, both formal and informal, sounded a general consensus. Although the prime target might be students, one must consider other audiences to reach the students. Without significant teacher involvement, the students will not be prepared for what you have to offer, or even value it. Without administrative support, the teacher workload is increased greatly, and whatever partnership you form will leave when the teacher leaves. The degree to which you focus on parent involvement depends on how well that is currently set up in the schools. If that is lacking in the schools, the success of your programs might depend heavily on drawing the parents in. Of course, home-schooled students have parent and teacher rolled into one, but the administrator takes on a different form. To form a lasting partnership, consider contacting the state or local home school association.

2. **Knowing what the schools want, how general or specific should you make your programs? What specific content should you address?**

First, the content issue. Obviously, you have a finite set of resources, but a little imagination, guided by the focus, purpose, and mission of your site can greatly expand what you offer. For example, you might have an exhibit detailing how oil is refined. Unfortunately, the relevant chemistry is not part of the standard curriculum in the local schools. Environmental issues, however, are high on their list. Build your program based on available resources; creativity comes in the approach and an interdisciplinary look at the curriculum and your site. One solution is to create a hands-on activity on the cleanup of oil spills that accompanies a tour through the exhibit. In this way, you satisfy the schools' needs to meet their standards while broadening the students' horizons with regard to oil production.

## Getting Started—Deciding What You Want

As stated in chapter 1, reflecting the content of the *National Science Education Standards* and your state or district standards is sure to strike a common bond between what the schools want and what you can offer.

> **Strike a Common Bond**
>
> During the last nine years, the education program has become an integral part of the park's ongoing operation. Schools use the programs because of two key factors, one of which is that all programs are tied directly to the Virginia Standards of Learning educational requirements and to National Science Education Standards. *Tim Taglauer, Shenandoah National Park*
>
> Maryland science standards and outcomes, as well as AAAS Benchmarks, are the foundation of our school programs. Wherever possible, they model behavior, competencies, and skills that Maryland students are expected to demonstrate on performance assessment tests given by the state. *Valerie Chase, National Aquarium in Baltimore*
>
> The activities in which students engage meet the national math and science standards and are considered "minds-on" and rich in math and science concepts. *Al Staropoli, ASPIRA*

A matrix of noncontent features of the *National Science Education Standards* is contained on page vi. And in case our message isn't clear, by finding a way to tie together what your site offers with the *Standards*, you are sure to create programs that will appeal to the formal educators in your area.

Which brings us back to the second question posed in this section. How general or specific should you make your programs? Is it enough to use the *National Science Education Standards*, or should you tie your programs to state standards or even district standards? Before answering that, realize that setting out to help the teachers in every state and district in your region meet their specific standards can be an overwhelming task for your staff, especially if the staff consists of you! The answer to this question will vary with institutions. Larger sites with separate education departments can obviously accomplish more than smaller sites. So, to a large degree, your answer will depend on your resources. Once you have decided on the scope of your program, however, there is a great way to ensure that your programs really do fit the needs of your intended audience: Involve teachers in the development of your programs. That's not always possible, but it seems to work best.

## Involve Teachers

Teacher involvement in the development of new programs creates a sense of ownership, gives the program credibility, and provides professional development opportunities.... The park is not developing something that is irrelevant or unusable. *Tim Taglauer, Shendandoah National Park*

Santa Monica Mountains National Recreation Area has great educators who have helped develop programs and materials. They willingly serve on advisory committees and have co-presented at conferences. *Barbara Applebaum, SAMO*

In several projects we have created a team of site staff and local teachers to develop new programs and field-test activities. We combined the classroom knowledge and expertise of the educators with the resource knowledge and informal education experience of the site staff to create programs that successfully integrate the classroom experience with the site experience. Through grants, we were able to pay stipends and/or provide for food, lodging, and expenses for the teachers. The teachers that participated in these projects demonstrated their commitment and desire to develop a quality program that would be available to other teachers as well. This involvement gave us credibility with the educational community and buy-in from teachers and administrators.

I have used two approaches in developing programs for schools. Each approach has advantages and disadvantages, but I have experienced success using each. The individual site or school situation should dictate which approach is best to use.

**a.** Develop programs in-house and then market what you create to the educational community. The site staff needs to do the necessary research to ensure the programs address the needs of the schools while meeting the needs of the site. This approach may need to be used if there are no educators available to help or there is a highly knowledgeable staff person.

**b.** Develop programs using a team of educators and site staff. The teachers and the site staff combine their knowledge and expertise to create programs that meet school and site objectives. This is probably the best approach to use when possible. *Tim Taglauer, Shenandoah National Park*

## Getting Started—Deciding What You Want

**3. How much, and what kind of, teacher preparation will you require? Will you offer training, and if so, how will you attract teachers to the training?**

Students who are prepared to learn are a whole lot easier to work with than those who aren't. Okay, so much for the obvious, but what does it mean to have prepared students? It is widely accepted among those who study learning that the human mind comprehends new material by tying that material to prior experiences and knowledge. In other words, we build upon what already exists in our brains. If the new material we're trying to comprehend doesn't connect well with what we already know, then we tend to reject the new material as irrelevant, or keep it around as an isolated, memorized bit of knowledge. Thus, a graduate-level seminar on theoretical physics isn't likely to mean anything to the average fifth-grader (or adult, for that matter), because he or she lacks the necessary structure (an understanding of the basic concepts of physics!) in which to place the incoming information. Getting back to our original question, we can say that a well prepared student is one who has the necessary background to comprehend what he or she will encounter at your site. More about that in the next chapter.

**Don't assume teachers will know what you want them to know and how to use your site.**

For now, let's make the not-so-large conceptual leap that if we want prepared students, we need prepared teachers. The teachers who bring students to your site must know enough to be able to prepare their students. At minimum, this means the teachers must be at least familiar with what your site has to offer. Students who at least know what they are about to experience and what they should look for are more likely to focus on the things you want them to.

For the students to have a richer experience, you might want the teachers to involve them in specific experiences prior to the visit. As not all teachers are comfortable with hands-on science, this means some kind of training for the teachers.

Some sites offer extensive workshops and trainings, while others limit their teacher training to providing pre- and post-visit materials. What your site does depends on the scope of what you want to accomplish in the students' visit. It is clear, however, that if you expect significant teacher training, you will have to offer some incentive, such as certification units or curriculum materials. The best teachers—the ones you want working with you—have many time commitments and won't waste time on endeavors they don't perceive as valuable.

## Entice Teachers to Attend Training

We require that teachers attend a workshop prior to bringing their students to the site for a field trip. At the workshop, we define the role and responsibility of the site staff and the educators. We provide instructions and materials to help the teachers understand our site, to use the programs, and to prepare their students for the learning experience. By attending the workshop, the teachers show their commitment and desire to make the best use of the programs that are offered, and they earn points for teacher re-certification. *Tim Taglauer, Shenandoah National Park*

To enhance our after-school science programs for students, we offer a three-day teacher training that focuses on learning how to use hands-on curricular materials. As an incentive, we offer a $200 stipend and free curricular materials of their choice at the end of the training. We have chosen the materials to supplement, rather than duplicate, what the teachers currently do in their classrooms. *Al Sturopoli, ASPIRA*

We conduct professional development for each of these programs at our district science centers on site at specific schools. Most of our trainings span two to five days. Some trainings are conducted at the sites where the program is conducted. Other times it's a combination of in-class and outdoor settings. We want to show teachers how many of the activities at these informal science venues can be simulated on their campuses. *Henry Ortiz, Los Angeles Unified School District*

PEEC, in partnership with NSTA, the National Park Foundation, and Delaware Water Gap National Recreation Area, supported by ExxonMobil, convened an institute for national park education specialists and teachers. The first of its kind in the nation, it incorporated the *National Science Education Standards* in school-based programs where National Parks as Classrooms were included in inquiry-based instructional programs. *John Padolino, PEEC*

### 4. Do you want a long- or short-term partnership?

This again, as in human relationships, depends on your situation. Are you just trying to get general public exposure for your site, or do you want some stability in an educational partnership and the recognition that goes with it? A few of our contributors have definite ideas on the subject.

# Getting Started—Deciding What You Want

## Decide What Kind of Partnership You Want

With the PARKS grant, we have been able to substantially expand our partnership efforts. We have created a set of pre/post test materials to assess student learning, we have designed and implemented kindergarten and first-grade programs that correlate with the programs for second- through sixth-grade students, and we have developed the "park portfolio" system to assess student understanding and appreciation over several years. The partnership has been awarded a Virginia Governor's Partnership in Education Award for 2000 as a successful example of a local school-business partnership. *Tim Taglauer, Shendandoah National Park*

Partnerships that are short term or based solely on grant funding die. We try to stick with Baltimore City Public Schools through thick and thin, sometimes having nice grant support to help them and sometimes working for free. In turn, they keep us in mind and ask us to help, even when we don't have big bucks and can only do small favors. *Valerie Chase, National Aquarium in Baltimore*

Los Angeles Unified School District (LAUSD) at the operational field level has been outstanding. They have provided meeting rooms and special equipment as needed; their coordinator has made himself available whenever needed. LAUSD has undergone major reorganization, and many people have been reassigned. Despite this, LAUSD has remained a strong, committed partner. They have applied for a new five-year National Science foundation Urban Initiative grant. This could pay for new technologies and additional teacher training. With this grant they plan to assist us with updating the Parks As Laboratories program and to develop additional programs. *Barbara Applebaum, SAMO*

Forming lasting partnerships requires a great deal of work and communication, but the rewards can be substantial.

## Questions Formal Educators Should Answer

### 1. What kind of site is right for your students?

The answer to this question lies in two factors: knowing your students and their experiences and knowing what educational opportunities various sites have to offer. If all of your students are likely to have visited the local zoo, that might or might not be a good place to establish a partnership. To find out, the best thing you can do is visit the zoo yourself and meet with the education department. Have them show you what would be involved in their program for your students. If it involves a guided tour of the animals and the reading of informational plaques, then the kids might have a blast but learn nothing. On the other hand, a guided tour could be very educational:

> If a teacher is working with a significant number of students from low socioeconomic backgrounds, these students might never have seen even the local museums, zoos, parks or other informal education sites. This makes local sites good experiences for these children.
>
> —Faimon Roberts

**Sign up for teacher workshops or other offerings from a museum—get to know it well.**

Knowing your students is key, but knowing the site is equally important. Your children might have been to that zoo a hundred times, but chances are they never got a behind-the-scenes look at zoo operations and procedures. If the zoo offers that, then you're in business.

Of course, this advice goes beyond zoos. Only by knowing what can benefit the students in your class, school, or district, and then knowing what various sites have to offer, can you find a good match. For best results, don't just talk to the educational people at informal sites; experience firsthand what they have to offer. That takes time and effort, but it beats a completely wasted field trip. The Parent-Teacher Organization doesn't have unlimited funds, you know.

### 2. Sites generally want partnerships with educators who need what they have to offer. Do you fit that description?

Again, the only way you will know the answer to this question is to visit the site. First, however, you need to know exactly how your current curriculum fits with what the school, district, or state requires. Sounds like a good time to form a committee! Find out where your curriculum gaps are, and then ask whether or not a site can fill the gaps or at least enhance what you already do. Perhaps they have subject area expertise that your school lacks. Perhaps they have access to facilities that would be cost-prohibitive should the school try to acquire them. The bottom line is that if a site really doesn't add to your students' experience in a substantive way, then it's a wasted field trip and certainly not a great candidate for a partnership.

#### Advice from the Field: Determine Whether the Site Will Enhance Student Learning

If the site has an education or information person, I get as much information and investigations as they have.... I need to know all of the activities that will be available on-site to prepare my students. I study the site and look for things that might be a hazard or that I might not want to let my students participate in (sic).

## Getting Started—Deciding What You Want

> I must assess student understanding about the site and make sure that visiting the site will add to the students' learning of a concept. I must make sure I feel the site can and will provide a safe and unique educational experience. Many times I have contacted sites and found that they are not really set up to do anything special for school groups.
>
> The best [formal education] partners are those who need help.
>
> Know what the site is about—goals, mission, and program offerings. Know how this relationship will be beneficial.

### 3. What should you expect from a partnership with an informal site?

First of all, don't expect an informal site to run the whole show. This is not an opportunity to drop off the kids and hope they learn something. What you can expect is for them to offer their expertise and resources and possibly work with you to tailor what they offer to your needs. If you expect that, however, be prepared to offer your time and resources to ensure that the goal is reached.

> **Advice from the Field: Manage Your Expectations**
>
> Take advantage of special programs offered by the institution, but don't expect a large institution to design something very specific just for your students.
>
> Don't ask a site to do something that isn't within their area of interest or expertise.
>
> Have a clear goal or idea to present to the site; be able to articulate what your needs and objectives are. Don't assume the site will be able to do this for you.
>
> Don't assume the site will be solely responsible for all aspects of the program or partnership. Be prepared to contribute time and effort in developing the programs or activities with the site.
>
> Don't assume the site is able to do everything you would like them to do; sites have budget and staff limitations, too.

Chapter 4

# Making It Happen

Okay, you're a formal or informal science educator ready to get out there and establish a meaningful partnership with your counterpart. How are you going to make it work? Well, the first order of business is the…

## Initial Contact

Although you might make your first contact at an educational conference or on a chance visit, initial contact usually starts with a phone call. If you're an informal educator, the call might go like this:

> Hello. I'm Wood Z. Owl, the director of education at Big Ball of Twine Regional Park, and we are interested in forming educational partnerships with schools in the area. We have a great plan that will benefit your school (class, district), and I'd like the opportunity to meet with you and discuss what we have.

At this point, you might just get a request to send brochures and other materials. That's okay, but after sending the materials, keep calling until you get a face-to-face meeting with someone. As stated in chapter 3, clear and open communication is essential, and that starts with actually meeting the people you will be working with. After the initial contact, phone and e-mail are great for ironing out details.

If you're a formal educator, you also need to make that first call and *set up a meeting*. Whether you're a classroom teacher or an administrator, it would be advantageous to meet *at the site*. In your office or classroom, you can talk about what the site has to offer; at the site, you can experience what the site has to offer.

As you progress from the initial contact, there are numerous signs that indicate a good partnership is forming, and numerous signs that a bad partnership is forming. The comments below might help you recognize those signs, and also serve as a model for how to, and how not to, present yourself to a future partner.

# Making it Happen

> ### Advice from the Field: Warning Signs and Good Omens
>
> I am tipped off to a potentially meaningful relationship when people reach out and sincerely express themselves to our students. I feel that things are going well when people recognize students for donations, contributions, and accomplishments.
>
> A warning signal to myself that may signal a potential problem is when people do not return my phone calls or e-mails. Potentially bad relationships usually result from lack of communication on all parties.
>
> One sign of a potential positive relationship is when the teacher or administrator approaches the site to ask how or what can be done to use the site's resources. A genuine interest, desire, and participation in looking for ways to use site resources to benefit the students is a good indicator of a positive relationship. Once such a relationship is in place, the effort to look for ways to improve and move the partnership forward indicates that things are going well.
>
> Another sign to look for is disinterest or excuses. We met with one school administrator to offer and describe a newly developed program. First, she was 30 minutes late for the meeting. Then, she routinely looked at her watch the entire time we were talking. Finally, she made several excuses for why her district probably wouldn't be able to participate.
>
> You know you have a good working relationship when teachers call you or e-mail you with ideas and suggestions and make changes to your materials and offer to share their modifications with others.
>
> Warning signs include teachers and schools not returning calls or letters, teachers sending substitutes to programs, and advisory committee members missing meetings and/or not providing input into the materials.
>
> We have never had a bad relationship with a school district or the State Department of Education. We ALWAYS have gone to administrators and asked about their needs and how we could work with them up front.

## Educational Materials

In chapter 3, we strongly recommended that the development of educational materials for a partnership be a joint effort between formal and informal educators. If you take that advice, then it's a good idea to have some sort of structure for their development (the materials, not the educators). There are numerous models people use for constructing educational activities, and if you're familiar with one particular one, then that should be your starting point. For those of you new to the process, we present here one model that is consistent with the recommendations of the *National Science Education Standards*.

- Set out in detail what you expect the students to learn from their visit to the site. What concepts should they understand? What facts should they remember? What activities should they be able to perform?
- Knowing what you expect the students to get out of their visit, ask what experiences they need prior to the visit to give them the best possible chance of understanding various concepts and procedures. Figure out how to provide these experiences and structure them into classroom activities. These are your pre-visit materials.
- Design the on-site activities and explanations so they build upon the pre-visit activities. Introduce new concepts so they connect the site exhibit with the students' prior experience.
- Ask what kinds of things the students should be able to do if they truly understood the various concepts and facts presented during the visit. In other words, how could they *apply* their knowledge to demonstrate understanding, while reinforcing the ideas they have learned? Structure post-visit activities so the students can demonstrate and reinforce their understanding. These should be entirely new activities that draw upon the concepts learned at the site.

Let's illustrate those four steps with an example. Suppose a site offers an exhibit, plus hands-on activities, that demonstrates what technology is and how the design process proceeds. Step 1 might result in the following list of outcomes:

- The students will know that virtually everything around them that is manufactured was developed through a process of design.
- The students will know the meaning of the terms "goal" and "constraint."
- The students will be able to identify the goals and constraints they use when designing something.

The formal language ("the students will…") isn't absolutely necessary, but it helps. What matters is that you clearly state what you expect the students to know or understand after the visit. The above list contains only three outcomes, but that's a lot to expect students to grasp in one visit. To increase their chance of understanding, it will help if they have some experiences to which they can tie their field trip (see chapter 3 regarding understanding and prior experience). That's where Step 2 comes in. What's important in pre-visit activities is not to teach the students exactly what they will learn at the site (e.g., explain to them the meaning of goals and constraints). Rather, you want to *set them up*, or prepare them, for what they will do at the site. With that in mind, pre-visit activities might include the following:

## Making it Happen

- Through discussion, and maybe a scavenger hunt, help the students understand the difference between natural and manufactured objects around them.
- Have the students create something that can perform a specific task. For example, they could use play dough to construct cups for two-year-olds, or make a paper airplane that can perform certain acrobatic feats. Without introducing the concepts of constraints or goals, have the students explain why they made the cup handle a certain size or why their planes had certain features. In this way, the students are *experiencing* the concepts of goal (the cup has to be usable by a two-year-old) and constraint (two-year-olds have small hands) without being introduced to the formal concept.

At this point, we should mention that pre-visit activities don't necessarily have to set up the learning of complex ideas. Perhaps the site will have all the necessary experience your students need. For example, if you're visiting a wetlands area, the first order of business upon visiting the site might be getting down and dirty and finding out what kind of life exists in the area. In this case, pre-visit materials might consist of discussing what the kids will be doing, safety procedures, proper behavior, and the like.

Now, on to Step 3 for the on-site materials. The key here is to draw on the students' prior experience, whether they got that experience in the classroom or as a first order of business on the visit. In our technology example, any introduction of the concepts of goals and constraints should refer not only to the site exhibits, but also to the pre-visit activities using cups and airplanes. A basic discussion of technology might begin with the students sharing their previously generated list of manufactured and natural objects. In other words, make the connection between what your site has to offer and what the students have already done.

Step 4 might not seem necessary, but to solidify the students' understanding, it's good for them to reflect on the visit and try to apply what they might, or might not, have learned. You could have them write an essay summarizing the visit, but that's about as fun and rewarding as "What I did on my summer vacation." Better to give them a chance to apply what they have learned in a new setting. Keeping with the technology example, you might ask the students to "use craft sticks and rubber bands to create a useful tool for the classroom." In doing so, they must clearly identify their goal (e.g., build a chalk holder so the teacher doesn't keep dropping and breaking the chalk) at the outset, and identify the constraints they are working under (e.g., I only have four craft sticks and the teacher is left-handed) as they proceed. This gives you a chance to see what the students gleaned from the visit, and also gives an opportunity to re-visit the basic concepts.

### Advice from the Field: Solidify Students' Understanding

All of the institutions have required teacher orientation programs that teachers attend before bringing students. Pre-trip materials are provided for the teacher to use in the classroom to prepare the students for the experience. Post-trip materials are also provided to reinforce and extend the experience.

The link between my science classes, the rest of the school, and our visit to an informal education site depends upon the teacher communicating to students and all other affected people why we are going to a site and my expectation about what we should learn while at that site.

The most important part of the field trip is what I have students do before we go on the field trip. I would like to have materials from a site that describe what students are going to see at different stops throughout the trip and what we should all learn by traveling to this site. At times I do special activities to help students learn specific background information that might be necessary to understand the experiences provided at the site. If students are not well prepared to understand how to use the exhibits and activities at a site as learning experiences, then the trip is not worthwhile.

Design a worksheet that makes students stop and really examine the collection to observe, record, and draw conclusions. Use large exhibits and then let the kids just look at others. Don't try to use the whole collection on a worksheet.

DON'T use treasure hunts—The students never look at the exhibits. Don't base worksheets on written explanations on the exhibits—the students can read a book in school.

Provide a well paced and curriculum-appropriate program. Don't provide a long lecture or a program for adults—that frustrates students with material they can't understand.

Plan a variety of activities for all learning modalities and for all ages.

## Teacher and Staff Training

This is an issue that is primarily a concern of the informal site. However, in evaluating a potential partnership, formal educators would do well to pay attention to how teacher and staff training are done by the site. Whenever possible, formal educators should be involved in the development and implementation of training.

## Making it Happen

Successful training depends first upon getting your audience. The motivation for site staff is clear—if they want a job, they'll go through training! For teachers, the motivation for training might not be clear. We discussed a few ideas for attracting teachers to training in chapter 3, so you might want to go back and check that out.

Assuming you have your teachers, here are a few time-tested guidelines for structuring the training.

> **When possible, involve educators in the design of programs and activities. Don't develop programs and activities in a vacuum; at least know and understand the needs and objectives of the educators.**

- Teachers trying to teach science without adequate subject-matter knowledge generally don't do a very good job of teaching. Therefore, make sure your training includes all the necessary background knowledge for people not only to present the prescribed program, but also to be able to intelligently respond to reasonable questions.

- Training should reflect the experience you want the students to have. If you have an inquiry-based, hands-on program, then don't lecture to staff and teachers about how to implement the program. Throughout your training, model the behavior you expect from your educators.

- Develop training materials using the same four-step process we described in the Educational Materials section above. Not only does that process result in training materials that promote understanding, it serves as a separate model for teachers and staff to understand how the site educational materials are structured.

- Make sure that the members of your staff who don't have previous classroom experience understand the needs of children and how to effectively control a large group. "On task" doesn't necessarily mean sitting quietly and listening. Good, interactive science often gets a bit noisy. On the other hand, a guide who has no control at all will have a difficult time helping students learn.

- Provide continual feedback to staff and teachers, even after a visit. Teachers need to know when they have brought an "under-prepared" class.

# Chapter 4

### Get Involved in the Development and Implementation of Training

Park staff, school district representatives, and university representatives meet and draft a proposed training agenda. Once the three sides agree, a proposal is submitted and approved. At first it was difficult, but once the relationships were built and people got used to working with each other, successive training approvals became easier. *Barbara Applebaum, SAMO*

We have *more than* 600 volunteers working in over 35 job descriptions…We recruit folks from many walks of life, going to churches and community organizations as well as advertising on TV. Volunteers who are exhibit guides attend six full days of training and take both written and performance tests. We have a special program for summer high school students who do the same training. *Valerie Chase, National Aquarium in Baltimore*

Work with educators to train volunteers and be certain that the volunteers understand young people and are not overly critical.

Teachers build on their own experience and expertise and through PEEC have the opportunity to learn in part from the collective wisdom and experience of colleagues and others. PEEC affords teachers the opportunities for collegial reflection. *John Padolino, PEEC*

## Administrative Concerns

Informal sites have many and varied administrative concerns, but as far as partnerships go, their main concerns are to help the formal educators with *their* administrative concerns. Therefore, we present here a list of issues formal educators have to contend with, along with "advice from the trenches" on how to make things run smoothly.

Field trips are a wonderful educational experience if the preparation is done before you leave the building.

### Covering for teachers

If a substitute is not available to cover classes not involved in the field trip, check schedules of other teachers in the department/school and ask them to host your students during their regular classes. Each period assigns five to eight students (with work) to different teachers. Another option is to get teachers to cover classes on their free periods.

**Making it Happen**

### Permission Slips

Parental permission is required, so prepare a standard permission letter on the computer that can be adapted for each trip. Send it out at least two weeks before the trip and expect it back one week before. Do not allow students without permission to go; if necessary, call home to get the form. Always have alternate students with permission ready to go if there are "no shows" on the day of the trip.

A letter and signed permission slip are necessary. The letter provides all information about the trip, giving details about where and why we are leaving the school. The slip signed by parents proves that I communicated with them. It is important to inform everyone of what you are trying to accomplish by leaving school. Make the letter and permission slip separate, so the information stays at home.

Go over all the details about the visit with your administrator ahead of time. Know how to abort the trip and who to contact and what to do in case of emergency.

### Chaperones

Chaperones are always needed—usually one per ten students. With the permission of the principal, use lab assistants, teacher aides, clerical staff, and parent volunteers. Invite the parents of students with the "carrot" that they do not have to pay. Talk with chaperones BEFORE the trip and have their duties clearly defined. Be certain they know their responsibilities.

Prepare your chaperones using materials provided by the institution. Don't ever use students as chaperones—even if they are 18—and don't let your chaperones bring their own small children on the trip.

### Transportation and Other Fees

Paying for transportation and other fees is always a problem. Some schools permit fundraisers and some have parent or community groups that can help. Small grants from local foundations are available in most areas. If a bus costs $136 for 40 students, then the per capita cost is $3.40 per student. To accommodate "free" chaperones and to provide a snack on the bus, or to buy a small souvenir of the trip for each student, charge $4.50 or $5.00. It is important to have the students contribute something, because they are more likely to take the experience seriously.

Make sure you know how the site expects to be paid for your visit and send the funds ahead of time (check to make sure they received the funds).

You do not want to travel and then be turned away. DO NOT carry cash to pay for your visit.

## Student Attire and Equipment

Some students are reluctant to participate in outdoor activities because they lack the proper clothing and equipment, and do not have funds to buy a lot of new things. Stress that new items are not needed. Be specific about what is needed. For an outdoor trip, for example, specify old tennis shoes that can get wet and muddy, two pair of old jeans, gloves that can get dirty, sunglasses, a sleeping bag OR blanket, a plastic cup, and no radios. Always take an extra supply of hats, gloves, jackets, etc. Indoor trips also require guidance, such as proper shoes, no hats, no shorts, no tank tops, etc. For lab experiences, specify no dangling jewelry, something to tie hair back, and protective gear such as goggles.

Describe what the students can and cannot bring— for example, no open containers of things to drink. I once left two dozen containers of orange juice upside down and emptying on the sidewalk because I thought they might contain liquor. Give specific instructions about what they should and should not wear. Students should not bring electronic devices such as video games, walkmans, or radios.

Ensure that students are prepared for the weather and are wearing appropriate shoes and clothing. Don't bring students out to a field program wearing their Sunday best clothing, sandals for hiking, or shorts and a tee shirt when it is cold and rainy.

## Student Behavior

Behavior expectations must be clearly defined AHEAD of time. Go over bus etiquette and how to act in public. Institutions have specifics, such as not feeding the animals, using elevators, or running. Remind students that they are representing themselves, their parents, their class, their school, their community, and you. Do not hesitate to make a student "sit it out" if he or she acts up during the trip. However, give all students a chance to participate. Often the class clown or troublemaker can be turned around by a field experience.

Give students an idea of acceptable and unacceptable behavior and the consequences for their actions. Tell site staff what you do not want your students to have access to, for safety or other reasons. Have a record of health concerns, medications, allergies, and such for each student.

## Making it Happen

# Chapter 5

# The Visit

> All the students were eager to get into the waders and try their skills at frog catching. We wanted this to be a learning experience, so we did not give a lot of instruction on frog-catching techniques.... Most hit the water as one herd, stomping and hooting. It did not take long for the students to realize this method was unsuccessful in frog catching, so they started to split off in quiet ones and twos. This was more productive. After about one hour we regrouped and talked about what had happened, who was successful, and why.... We caught and identified turtles. We observed baby turtles the size of 50-cent pieces. We cut into old pitcher plants and observed the remains of insects. We followed bear tracks in the sand. We watched insects molt and found millions of tadpoles...The learning curve for these students was steep. They were all actively engaged and involved in the study. This was an example of what could happen for students in science education. The experience was in-depth and relevant and involved application of the scientific mode of problem solving. Students were absolutely unaware of the ecology of the wetlands on the islands and had very little field experience in traditional science classes. Students learned firsthand that field research is hard and sometimes time consuming, but rewarding work. You never know what you might find. We all had a wonderfully enriching experience.
>
> —*Educator at Apostle Islands National Lakeshore*

Well, that sounds great, doesn't it? If every field trip you're involved in has that flavor, then whether you're the site or the teacher, you're in business. Such experiences are entirely possible, but they don't happen without a bit of planning and a lot of common sense. In this chapter, we're going to give you some strong recommendations about how to complete a successful field experience, accompanied by true anecdotes and a bunch of do's and don'ts suggested by our contributors.

## Recommendations for Formal Educators

### Participation

The last thing an informal site wants you to do is drop the kids off and go have a cup of coffee. They're not baby-sitters. The whole point of a field experience is for you

# The Visit

> **Participate in training, workshops, and the programs with the students. Don't expect the site to be your substitute teacher.**

and the students to expand beyond your classroom and enhance your learning, so get in there and get "dirty" along with the students. Ask probing questions that help move the discussion along, and assume an active teacher role when it's appropriate, such as moving around to help small groups accomplish a task. That said, this is not a time to show off your vast knowledge about the subject at hand. Be careful not to "upstage" the site staff, or to start a discussion the staff wants to postpone just a bit. Also, keep your discussions with the site staff on topic—yes, that was a great game last night, but this isn't the time to analyze it.

Not actively participating sends a message to the students and to the informal site that what they're doing isn't important to you. Not a good way to start a partnership with the site that you expect to last for a number of years. Also, if the educational materials are developed in a thoughtful way, there will be post-visit activities for you to do with the class back at school. It's unlikely that will go well if you have no idea what the kids did at the site. Most sites will welcome the active participation of teachers and chaperones. If not, rethink that partnership idea.

## Safety

Because you have already visited the site prior to the field experience (right?), you know about possible safety concerns during the trip. You know what areas the students should avoid, what exhibits represent potential hazards, and how to keep from getting lost. Your advance visit is especially important in the case of manufacturing plants or other sites that might not have a long track record of accommodating large groups of people.

Armed with your safety information, you really ought to share it. So, before you even leave the school, it's time for a safety lecture. Discuss the above, plus what to do if an injury occurs, what to do if a student *does* get lost, and the like. When you reach the site, and prior to getting off the bus, run through the safety issues again. After the students are off the bus and have a chance to use the restroom—you guessed it—run through the safety issues again. The site staff might do this last run-through; discuss this with them ahead of time. Also discuss any potential medical problems with the site staff ahead of time. Most guides would appreciate knowing whether any students are allergic to bee stings as they head into the Making of Honey exhibit.

Don't assume the site will have first-aid kits—carry your own. Check the contents of the kit before you leave.

## Chapter 5

### Keeping track of students

> One school group got on the bus and drove away without a chaperone and four kids. We expect the teacher was not too well received when she got back—we called the school to tell them their lost folks were with us.

It is generally considered bad form to return to school with more or fewer students than you started with. So, begin by having a complete list of all the students, teachers, and chaperones who are going on the trip. Include phone numbers and any necessary medical information, such as allergies, on this list. Take roll before you leave the school and before you get off the bus at the site. Take roll again at lunch or during any other break, and certainly take roll before *any* of your crew leaves the site. It's a nice courtesy to give a copy of your roll sheet to a site representative.

Provide name tags, especially for younger students. Middle and high school students might balk at the name tags, seeing as how they are spending a lot of time being cool and trying to be cool, and—let's face it—students don't think name tags are cool. However, it is sometimes nice for site staff, and your own parent chaperones, to be able to call students by name. One way to avoid the potential lack of coolness, is to let the students design their own name tags or incorporate a twist based on the story of the site.

### Discipline

The quickest way to be "dis-invited" from further participation with a site is to show up with unruly students and expect the site staff to deal with the problem. Yes, the site staff should exhibit some level of control over the group, but they shouldn't be expected to handle extraordinary behavior. Your first order of business along these lines should be to train any nonteacher chaperones with respect to keeping the kids in line. Tell them what behaviors will and will not be tolerated, and have clearly defined consequences for violations. You can't send children to the principal's office once you're on-site, so have a place at the site, or possibly on the bus, where you can send them to a supervised detention. It might be a good idea to have work available to keep their brains going.

During the trip, don't be afraid to step in and take care of any potential or occurring discipline problems. It is a rare guide who will resent that kind of intrusion. By the same token, if one of the site staff issues a disciplinary warning to one of your students, add your approving look for reinforcement.

### General concerns

Remember that you and your students are guests at the site. First, don't bring any surprises with you. Show up with the agreed-upon number of students and

# The Visit

chaperones. Don't bring pets or chaperones' children unless specifically told that it's okay to do so. If you have unavoidable last-minute changes, communicate those to the site *before* you leave the school.

> A minority of teachers have canceled programs at the last minute or not shown up. One teacher brought an unleashed, hyperactive pet with her; and in one instance we had a student try to set fire to a park structure while the teacher was looking. The worst incident was a teacher who let her students wander off at lunch and didn't remember how many she brought with her to the park.

Be on time and be polite. Come prepared both physically and mentally. Make sure you *and* your students show respect and pay attention. After the visit, send thanks to all involved. Drawings from little ones, and sincere notes from older students, are always welcome. Note, however, that whatever you send to the site reflects upon your school. It would be a good idea to screen whatever material you send. Now, although site personnel love to get thank-you's, they cringe at 60 identical requests for further information or brochures. Make sure whatever you send post-visit doesn't require a dramatic increase in the site's workload.

### Recommendations from Informal Science Educators

Do not bring too many or too few students. Arrive and leave on time. Do not expect special consideration and do not change plans.

Teachers who expect the staff to maintain discipline and regard a visit as a day off aren't well loved.

Take care of discipline so the guide can present the program. Don't ignore disruptive students.

After any visit send personal thank-you's to anyone who facilitated your success.

We have seen that some schools are discouraged from participating in our programs because of our policies and guidelines. For instance, we require that teachers attend a workshop prior to bringing their students on a field trip, and we limit the number of students to 60 per program so that we can maintain a ratio of one ranger to 15 students. We stress the quality of the experience over trying to accommodate as many students as possible. Some schools are required to bring the entire grade level on field trips—if this is more than 60 students, the school must come on multiple days or not participate in the ranger-led program.

NEVER ask all the students in your class to write us for information about

> the same things all at once. Don't expect us to be able to have students shadowing our staff—we get hundreds of requests and cannot do it.
>
> Allow students to have fun while connecting with the resource.

## Recommendations for Informal Sites

The major transportation issue at our park is the travel distance from the schools. In many cases the travel time is one and one-half to two hours each way. We have tried to address this issue by selecting new program sites that are closer to the park entrances. However, we must be careful to allow groups the opportunity to visit the restroom facilities before and after their program.

### *Facilities*

Whatever kind of field experience you have to offer, no one will want a return visit unless you have adequate and clean facilities. Make sure you have enough space and enough staff to accommodate whatever size groups you have visiting. Consider scheduling school groups during nonpeak hours and nonpeak seasons.

Restrooms are a major priority when 30 or more students visit. Figure out, logistically, when restrooms will be available and when they won't, and inform the school personnel. Start the trip with a restroom break and let people know when they'll get another one. Rustic restrooms are fine; dirty ones aren't.

> **Provide clean restrooms and allow the students to use them!**

Communicate with the school ahead of time to see if they have any students with special needs. Make sure you have adequate access for physically challenged students. If not, inform the school as to what such students will and will not be able to do on the trip. As with all phases of such partnerships, keeping open lines of communication will solve most problems before they occur.

If you are an outdoor site, have alternate plans in case of bad weather. If you simply don't have a protected area where students can do meaningful activities, arrange a cancellation plan with the school. A bus or two of hyperactive kids waiting out bad weather is not the ideal field trip.

### *Safety*

Once I took my Environmental Club members on a cold Saturday morning to a stream bank restoration planting along an exclusive golf course. An

# The Visit

> uninformed homeowner/golfer proceeded to verbally abuse a student who walked across the course. Before I interfered, he was actually waving his golf club under her chin.... I think our informal education connection could have been more effective in the public relations department. The homeowners should have been made better aware of how our students were volunteering and why.
>
> —Sheila Hodges

Assess your facilities for any potential safety hazards and whether those hazards apply only to younger students. Communicate this information to the schools. Make a discussion of safety your first order of business when the students get to the site. Outline what to do in case of injury or other emergency. It would be a good idea to work with the formal educators on developing the safety discussion so the students aren't getting conflicting messages.

All members of your staff who will be working with students should have first aid and CPR training. Also, don't depend on the school staff to have their own first-aid kits. You should have adequate kits available at all times during the field experience. Having to "run back to the lodge" that's a mile away could mean compounded problems and a possible lawsuit.

If your site is not isolated, or you arrange excursions into populated areas, make sure you have cleared this with all of the parties involved. Unexpected encounters can be a safety issue for students.

## *Discipline*

> One middle school brought the entire school (before we learned to make rules about group size) and let them run wild. Kids were throwing things from one level to the next, etc. It was insane, and they were banned from visiting for more than five years before we gave them another chance.

In the ideal situation, any school groups that visit your site will have adequate supervision from an adequate number of chaperones. This not being an ideal world, you should be prepared. First of all, make sure all of your staff know how to deal with children. Previous experience in the classroom is not a guarantee! There are basic techniques for keeping things under control, and good teachers use them with seeming ease. Therefore, one of the best things you could do is find an area teacher who excels in this area and arrange for him or her to meet with your staff. There is also no substitute for observing such a teacher in action. With the administration's permission, arrange to have your staff sit in on this person's classroom for a few days.

Even with a well trained staff, you will run into problem kids. Hopefully, the chaperones will deal with any bad situations, but just in case, you should develop clearly defined consequences for disruptive behavior and be ready to administer them. It's a good idea to have an area set aside where disruptive students can go and be supervised.

Of course, the best way to avoid discipline problems is to keep the students actively engaged. If you require the students to be absolutely quiet, have a "hands off" policy on all exhibits, and talk *to* the students instead of talk *with* them, you're asking for trouble. Active minds tend to spend less time trying to cause trouble. Also, remember that a field experience should be fun as well as informative. Let the kids be kids. And don't be completely rigid in your scheduling. If things will run more smoothly by doing activities in a slightly different order than you planned, then by all means do so.

> **Even if you have tons of rules, try to be flexible when you can. Give a member of your staff who has seen it all the freedom to choose when to break the rules.**

# The Visit

# Chapter 6

# Maintaining the Relationship

## Key Ingredients

When students experience a project—whether it is designing, constructing, planting, and maintaining a wildlife habitat; monitoring a weather station; tagging black bears; stocking trout; climbing fire towers; or restoring pristine trout streams—they visually express their sense of pride and accomplishment. Then I know I have accomplished my mission! Educational partnerships work.

—*Sheila Hodges*

It's unlikely that formal and informal educators get together for coffee, talk over a few options, and decide then and there that they have a lasting partnership. Things build slowly. But what are the key ingredients for creating a lasting relationship? How is it that a science education site and a school or district keep things going year after year? Chances are, you already know what many of those ingredients are:

1. **Common goals that benefit both partners.** Each side has to want the same end result, and must derive something positive from the collaboration. Generally, the end result both sites and formal educators look for is students who are excited about science, see it as a process of inquiry, and realize that learning about science is not confined to textbooks. The benefits for each side, other than the satisfaction of reaching the goal and knowing that everyone involved is just pretty nice and wonderful for reaching the goal, might be quite different. For formal educators, the major benefit might be an established avenue for helping the school or district meet its science requirements that are consistent with, say, the *National Science Education Standards*. For the site, the major benefit might be increased exposure and publicity in the community, which translates into increased overall attendance and further dissemination of the site's mission. As we've stated in earlier chapters, a long-standing partnership also makes everyone's life easier in that both sides know what to expect when students visit the site.

As we saw in some of the anecdotes in chapter 5, stability and knowing what to expect can be a very good thing!

2. **Give and take.** Basically, it's a good idea for site staff and formal educators to help each other out through volunteer efforts. Site staff can help with science fairs, serve on curriculum committees, and help in teacher training. Formal educators can volunteer at the sites and serve on site advisory committees. This keeps the connection going, while helping each partner better understand the capabilities and limitations of the other.

> Educators who volunteer as exhibit guides learn more, know more about how to use us, and get more freedom to do special things with their students. They can borrow things from us, use our classrooms, and do things most school groups cannot because we don't have enough staff.
>
> —*Valerie Chase, National Aquarium in Baltimore*

There will be times in a long-term relationship when one party or the other is a bit less able to pull its weight. This will usually center on money and the ability to fund the field experiences, but can also have to do with human resources, such as when a key liaison leaves either the site or the formal education establishment. That's when the other party needs to pick up some slack, be it with a grant or by helping to train new personnel. We repeat here a quote from the National Aquarium in Baltimore.

> We try to stick with Baltimore City Public Schools through thick and thin, sometimes having nice grant support to help them and sometimes working for free. In turn, they keep us in mind and ask us to help, even when we don't have big bucks and can only do small favors.
>
> —*Valerie Chase, National Aquarium in Baltimore*

3. **Communication.** Every contributor to this book has stressed, at one point or another, the need for ongoing and constructive communication between the parties. This applies to small issues, such as keeping each other informed of small changes in plans, and large issues, such as suggestions for changes in the curriculum. Contrary to what many formal educators might think, suggestions for change and constructive criticism are most welcome at the sites.

### Develop a Common Mission

One of the keys to the success of this partnership has been the synergy and relationship that has been developed between the school principal, the education director from Shenandoah National Park, and the representative from Merck, Inc. At the core of this is the common mission and goal statement that was developed during the planning year/phase of the partnership.

These three people, along with teachers and the McGaheysville's assistant principal, meet formally four times a year (twice in the fall and twice in the spring), at meetings of the Business Partnership Steering Committee. This group is responsible for planning, oversight, and evaluation of the activities related to the mission and goals statements. These meetings provide the formal means for communication. Summary notes from the meetings are shared with the faculty and staff, as well as with the members of the McGaheysville School Council (a decision-making group of teachers, staff and parents).

Other vital communication happens informally and regularly throughout the year, however. E-mails and phone calls to share ideas and grant possibilities, for example. Summer lunches — when I can get away from my building for a bit — to talk about how things are going and to brainstorm where things might head. *David Burchfield, McGaheysville Elementary School*

4. **Evaluation.** A relationship that lasts is one that you continually evaluate. Find out what's working and keep it. Find out what's not working and change it. Easily said, but what should you evaluate and how should you do it? We'll use the PARKS model of evaluating partnerships as a guide. Though the specific subject of evaluation will vary from site to site, and how that evaluation is conducted will also vary, there are general guidelines that cut across all disciplines and facilities.

It's not our intention to get into a whole bunch of jargon about evaluation, but there is one distinction we have to make, and that is the difference between the evaluation of **cognitive** measures and **affective** measures. Cognitive measures refer to what someone knows, understands, or is able to do. For example, if you have just had a field trip to a newt museum, you might evaluate whether or not your students possess the cognitive skill of being able to distinguish a newt from a tadpole (note that we're at a low level of comprehension, here!). Affective measures, on the other hand, refer to people's attitudes, values, or feelings about a subject. For example, you might measure whether your students' excitement about doing science, especially using newts, has increased or decreased.

> Teachers who give us the toughest, most thorough evaluations are asked to serve on an advisory committee.

Part of evaluating a partnership is identifying various cognitive and affective measures, and asking whether the people involved are performing as you would like on these measures. Some examples should clear this up. In fact, we'll give you more than examples. Below are a relatively complete set of cognitive and affective questions

**Maintaining the Relationship**

you might want to ask regarding three of the main players in the partnership: teachers, students, and site staff.

### Teacher Affective Questions

- Are the teachers involved positive about the current partnership, and do they want to continue this partnership and pursue others?
- What are the teachers' attitudes towards informal science educators as valuable or not-so-valuable resources?
- To what extent do the teachers think the partnership is helping them reach various goals, such as meeting state science standards?
- How do the teachers compare, in terms of value, time spent at the site with time spent in the classroom?

### Teacher Cognitive Questions

- How well do the teachers understand the various concepts presented at the site?
- To what degree do the teachers follow the pre- and post-visit curriculum materials?
- To what degree do the teachers connect the field experience to their regular classroom activities?
- Can the teachers articulate their role in the partnership and compare it with the role of site staff?

### Student Affective Questions

- Are the students enthusiastic about the site visits?
- Have the site visits increased or decreased the students' enthusiasm for science in general?
- Have the site visits changed the students' awareness of, and concern for, environmental or other issues that are the focus of the site?
- How do the students compare, in terms of value, time spent at the site and time spent in the classroom?

### Student Cognitive Questions

- After a visit and doing post-visit activities, how well do the students understand the various concepts presented at the site?
- After a visit and doing post-visit activities, how well do the students perform various procedures or tasks presented at the site?

- To what extent are the students able to identify connections between material presented at the site and their regular curriculum?

### Site Staff Affective Questions

- Are the site staff members positive about the current partnership, and do they want to continue this partnership and pursue others?
- What are the site staff members' attitudes towards having school groups visit the site?
- To what extent do the site staff members feel that the formal educators support what the staff is trying to do, both in deed and attitude?
- To what extent do the site staff members feel the formal educators value them as a resource?

### Site Staff Cognitive Questions

- Do the site staff members understand how the content presented at the site connects with the school or district science standards?
- Are the site staff members able to communicate necessary concepts effectively?
- How well do the site staff members tie site concepts to the pre-visit activities?
- Do the site staff members model an inquiry approach to learning science?

Before we continue with other kinds of questions you might want to ask regarding how well the partnership is going, stop a minute and ask *how* you're going to get the answers to the questions presented so far. The answer is, not without a lot of work! Asking the right questions is only the start. A lot of the affective questions can be answered using interviews and surveys, but there's an art to developing such things. Answering cognitive questions can be even more difficult. If you have an evaluation expert at the site or at the school, great. Otherwise, you'll have to look elsewhere. One place to start is at the district or state level of the school system. Find such a person, procure a small grant to pay him or her, and you're in business. Another place to look is at a nearby college or university. Any department of education will have an evaluation expert. And if you have an existing partnership, adding university personnel to the picture will greatly enhance your ability to get a bit of money for an evaluation project.

Now that you know evaluation is no small task, here are a few more things you might want to evaluate.

### Program Materials

Teachers at McGaheysville, along with SNP education staff, have been directly involved in the development of the pre- and post-visit activities and assessments. The pre- and post-tests are the same instrument, in a multiple choice format, designed to measure knowledge before and after the study. The tests were revised after the first year of use, based on feedback from teachers at McGaheysville, as well as from other educators that used them.

Perhaps the most interesting idea concerning assessment that has emerged from the partnership has been the use of "Park Portfolios." The portfolios are records of student learning over the years they are at McGaheysville. They include pre- and post-tests, student drawing and writing samples, and pictures and descriptions of projects students have completed to demonstrate their understandings.

—David Burchfield

- Do pre- and post-visit materials enhance, inhibit, or have no effect on the students' cognitive measures with respect to the park?
- Do teachers and site staff find the program materials easy or difficult to understand and use?
- To what extent do the materials reflect district, state, or national science standards?
- Do the materials promote an inquiry approach to science?
- Are the materials developmentally appropriate with respect to both knowledge and skills?
- To what extent have the site's materials changed as a result of the partnership?

## *Communication*

- Do site staff and formal educators feel that there are open lines of communication between the two?
- What structures (e.g., committees and regularly scheduled meetings), if any, are in place to ensure long-term interaction between formal and informal educators?
- Is decision making equally divided, or does one member in the partnership have significantly more power to make decisions?

You get the idea, and you can most certainly add to the list and customize the list to fit your partnership. The main point we're trying to make is that continual evaluation can do nothing but improve the partnership over time.

# Chapter 6

## Advice from the Field: Notes on Evaluation

We are using a PARKS grant to develop, pilot, and produce comprehensive evaluation materials including student and teacher program evaluations and attitude measurements. We are currently in the pilot stage…Some of the PARKS monies were used to have a NSTA evaluator observe and review our programming. In addition, the park uses the Assessing Parks As Classrooms: A Model for Program Evaluation to measure our education programs' effectiveness. This system employs unbiased, trained National Park Service personnel from outside the park to conduct evaluations.

Through evaluation and follow-up, we work closely with teachers to ensure that the programs meet the state standards.

We do not do assessment of student performance, leaving that up to the teachers. We do formative evaluation of our programs to make sure students are learning what we think they are. We provide a simple assessment tool on the evaluation form the teachers are asked to return to check for learning along the way.

Provide constructive feedback, especially when asked. Look for ways to improve the relationship, and don't neglect evaluation forms. Approach a site with an idea—they might have never thought of it.

Evaluate, Evaluate, Evaluate! Communicate with educators and ask for constructive feedback. Revise and adjust as needed for continuous improvement.

Assessing program effectiveness is very important to us. We have a pre-test/post-test system for assessing acquired knowledge. We depend on the teachers to report this information to us. During our first year, the results of the reported scores indicated improvement for all programs.

Comments on the program evaluations have shown that because the programs are directly tied to the *Standards*, teachers can easily justify the time, effort, and expense of a park visit.

Ask for student and teacher evaluation and follow through on suggestions for improvement. Don't be defensive about comments.

We always like getting copies of worksheets, from teachers and appreciate when they take the time to thoughtfully fill out the evaluation forms.

Do serious formal evaluations on your programs and workshops. Use it to make your products and services better and publish your results to help others.

## Maintaining the Relationship

*Funding.*

Funding is provided by many agencies. We use private grant funds, Eisenhower funds, or school district funding for the different sections of the program. Therefore, some funds can be used for certain facets of our programs and not for others. For some programs, we require the school to provide the funding for 50% of the program costs. For example, we provide two substitute days and the school provides two days. Up to this moment, our school district administrators have realized how beneficial these programs are for students and have provided funding for the students.

—*Henry Ortiz, Los Angeles Unified School District*

> **The partnership helps with grant funding, lobbying local and state governments for support, and appeals to private funders.**

Maybe about now you're wondering how in the world you're going to pay for all of this. Why, have a bake sale, of course! Actually, we're only half kidding. School fundraisers and contributions from parent-teacher organizations can help a lot. Beyond that, the most precious resource you have is the partnership itself. Funding agencies just love it when different kinds of groups get together for a common cause. Check into local grants for school-business partnerships. Bring a university into the partnership, with people who are used to applying for grants, and you're in even better shape. At the national level, there are a large number of agencies that fund partnership activities, including the National Science Foundation, the Department of Education, and various corporate entities. We have included several websites of funding sources among the web resources listed on page 55.

# Web Resources

### AAAS

The American Association for the Advancement of Science (AAAS, pronounced "Triple-A-S") is the world's largest general science organization and publisher of the peer-reviewed journal *Science*. With more than 138,000 members and 275 affiliated societies, AAAS serves as an authoritative source for information on the latest developments in science and bridges gaps among scientists, policy-makers and the public to advance science and science education.

*www.aaas.org/*

### AABGA: American Association of Botanical Gardens and Arboreta

AABGA's mission is to support North American botanical gardens and arboreta in fulfilling their missions to study, display, and conserve living plant collections for public benefit.

*www.aabga.org/*

### AAM: American Association of Museums

The American Association of Museums is the national organization representing the museum community and addressing its needs, thereby enhancing the ability of museums to serve the public interest.

*www.aam-us.org/index.htm*

### AZA: American Zoo and Aquarium Association

Established in 1924, the American Zoo and Aquarium Association (AZA) is a professional organization dedicated to the advancement of North American zoos and aquariums through conservation, education, scientific studies, and recreation.

*www.aza.org/*

# Web Resources

## ASPIRA

The ASPIRA Association promotes the empowerment of the Puerto Rican and Latino community by developing and nurturing the leadership, intellectual, and cultural potential of its youth so that they may contribute their skills and dedication to the fullest development of the Puerto Rican and Latino community everywhere.

*www.aspira.org/*

## ASTC: Association of Science-Technology Centers, Incorporated

The Association of Science-Technology Centers, Incorporated (ASTC), is an organization of science centers and museums dedicated to furthering the public understanding of science. ASTC encourages excellence and innovation in informal science learning by serving and linking its members worldwide and advancing their common goals.

*www.astc.org/*

## The Foundation Center

The Foundation Center's mission is to support and improve institutional philanthropy by promoting public understanding of the field and helping grantseekers succeed.

*www.fdncenter.org/*

## IMLS: Institute of Museum and Library Services

The Institute of Museum and Library Services (IMLS) is an independent federal grant-making agency that fosters leadership, innovation, and a lifetime of learning by supporting museums and libraries.

*www.imls.gov/*

## Louisiana Public Broadcasting

Education has always been the focal point of LPB's 20-year mission. LPB has produced a long list of award-winning educational programs including *The Power of Algebra*, *Savalot's Energy Secrets*, and *Dr. Dad's Ph3* in addition to numerous other in-service programs and college credit courses for educators. LPB also houses the *Louisiana Educational Technology Resource Center*, which assists teachers and administrators with questions about implementing technology in their schools.

*www.lpb.org/*

# Web Resources

## National Aquarium in Baltimore

The National Aquarium in Baltimore offers a wide variety of programs for students and support for teachers both during their visit to the Aquarium and in the classroom. These programs include teacher-led tours of the Aquarium, private classrooms where students can interact with animals, auditorium presentations, and more.

*www.aqua.org/*

## National Park Service

The National Park Service preserves unimpaired the natural and cultural resources and values of the national park system for the enjoyment, education, and inspiration of this and future generations. The Park Service cooperates with partners to extend the benefits of natural and cultural resource conservation and outdoor recreation throughout this country and the world.

*www.nps.gov/*

## National Science Foundation—Funding

The NSF funds research and education in science and engineering, through grants, contracts, and cooperative agreements. The Foundation accounts for about 20 percent of federal support to academic institutions for basic research.

*www.nsf.gov/home/grants.htm*

## PEEC: Pocono Environmental Education Center

The Pocono Environmental Education Center (PEEC) advances environmental awareness, knowledge, and skills through education in order that people may better understand the complexity of earth systems. PEEC serves as a residential center for environmental studies; convenes institutes, workshops, and symposia for formal and informal educators; and develops and implements education programs for use in kindergarten through high school, college and university, as well as adult and continuing education.

*www.peec.org/*

## Santa Monica Mountains National Recreation Area

Located in the Los Angeles and Ventura Counties, Santa Monica Mountains National Recreation Area (SAMO) provides *Standards*-based programs for teachers and students to learn about urban-wildland interactions; ecosystems; plants and animals; and the effects of urban areas on water, air quality, and biodiversity. For more than 10 years, SAMO has worked in close partnership with the Los Angeles Unified School District.

*www.nps.gov/samo/educate/educate.htm*

## SchoolGrants

The purpose of *SchoolGrants* is to provide resources for children, teachers/educators, and K–12 schools. The goal is to contribute to the common task of improving schools and programs for the children served by those schools.

*www.schoolgrants.org/*

## Shenandoah National Park

Shenandoah National Park strives to provide education programs that address the needs of schools and neighboring communities. Education programs employ hands-on, sensory-based activities and encourage problem solving and critical thinking. They communicate the mission of the National Park Service and the significance of Shenandoah National Park, and foster environmental stewardship. All programs support the Virginia Standards of Learning.

*www.nps.gov/shen/*

## U.S. Department of Education

This Web site lists funding opportunities and application procedures for federal grants, programs, and awards.

*www.ed.gov/funding.html*

## The Why Files

The Why Files uses news and current events as a springboard to explore science and the larger issues it raises. It also shows science as a human enterprise and a way of looking at the world. Beyond portraying the outcomes of science, the overarching goal of this Web site is to explain the process, culture, and people that shape science.

*whyfiles.org/teach/index.html*

# Appendix A:
## About Parks As Resources for Knowledge in Science (PARKS)

### PARKS Program Description

Parks As Resources for Knowledge in Science (PARKS) is a partnership of the National Park Service, the National Park Foundation, the National Science Teachers Association, Ohio State University, and ExxonMobil. PARKS supports science education reform through integrating National Park education programs with the *National Science Education Standards*. The program is also designed to build strong bonds between formal and non-formal educators, supporting both science education and stewardship for park resources.

### PARKS Goals:

- Integrate the *National Science Education Standards* into National Park curriculum-based education programs.
- Create a model framework for integrating national education standards into other park programs (for example, history or social studies programs).
- Promote the National Parks as learning laboratories and provide opportunities for students and teachers to use National Park resources to achieve the *National Science Education Standards*.
- Enhance the quality of science education for students nationwide.
- Increase knowledge of, and stewardship for, National Park resources, the National Park System and associated resources.
- Establish Parks As Classrooms® (National Park Service education program) models that can be replicated at other national parks, while fostering the incorporation of the *National Science Education Standards* into existing and future National Park Service education programs.

Thirty-two parks were chosen, through a competitive process, to receive $25,000 grants over a two-year period and send a teacher/ranger team to a *National Science Education Standards* workshop. Four additional parks received $10,000 grants and served as workshop hosts-sites.

## Appendix A

Park rangers are working collaboratively with local school systems and teachers to integrate the *Standards* into existing park curriculum-based programs and, in many cases, create new *Standards*-based programs. The National Park Foundation coordinates the program and serves as the fiscal agent for the parks. The National Science Teachers Association provides professional expertise on the *National Science Education Standards*. Ohio State University is evaluating the program on a national level and assisting park teams in designing and conducting local program evaluations.

# Appendix B:
## Directory of PARKS Participants

The following 36 facilities were selected from a system of over 380 natural and cultural areas; grantees included a mix of both categories. The information below was gathered from the National Park Service "ParkNet" Web site (*www.nps.gov*) and was supplemented by program data from the Parks As Resources for Knowledge in Science Annual Report, 1999. The ParkNet links—including Park descriptions, contact information, and hours of operation—are current as of May 2001.

### Apostle Islands National Lakeshore

| | |
|---|---|
| PARKS Program: | "Lake Superior Studies" |
| Program Partner(s): | High school teachers in Washburn, Wisconsin |
| Subjects of Study: | Wetlands ecology |
| Address: | Headquarters Visitor Center<br>Route 1, Box 4<br>Bayfield, WI 54814 |
| Phone: | 715-779-3397 |
| Email: | APIS_Webmaster@nps.gov |
| URL: | *www.nps.gov/apis/* |

**National Park Service Description**
Wisconsin's northernmost landscape juts out into Lake Superior as the scenic archipelago known as the Apostle Islands. The national lakeshore includes 21 islands and 12 miles of mainland Lake Superior shoreline, featuring pristine stretches of sand beach, spectacular sea caves, remnant old-growth forests, resident bald eagles and black bears, and the largest collection of lighthouses anywhere in the National Park System.

**Operating Hours**
The lakeshore is open all year. The headquarters visitor center is open daily from May through October from 8 a.m. to 5 p.m. (8 a.m. to 6 p.m. Memorial Day to Labor Day) and Monday to Friday November through April from 8 a.m. to 4:30 p.m.; closed Thanksgiving Day, December 25, and January 1.

## Appendix B

### Badlands National Park

| | |
|---|---|
| PARKS Program: | "Evolving Prairie Series" |
| Program Partner(s): | K–6 teachers and educators from Pine Ridge Indian Reservation, Interior, South Dakota |
| Subjects of Study: | Prairie ecology, geology, paleontology, human history, and resources management/stewardship of the land |
| Address: | Badlands National Park<br>PO Box 6<br>Interior, SD 57750 |
| Phone: | 605-433-5361 |
| Fax: | 605-433-5404 |
| Email: | badl_information@nps.gov |
| URL: | www.nps.gov/badl/ |

**National Park Service Description**

Located in southwestern South Dakota, Badlands National Park consists of acres of sharply eroded buttes, pinnacles and spires blended with the largest, protected mixed grass prairie in the United States. The Badlands Wilderness Area covers 64,000 acres and is the site of the reintroduction of the black-footed ferret, the most endangered land mammal in North America. The Stronghold Unit is co-managed with the Oglala Sioux Tribe and includes sites of 1890s Ghost Dances. Established as Badlands National Monument in 1939, the area was redesignated "National Park" in 1978. Over 11,000 years of human history pale to the ages' old paleontological resources. Badlands National Park contains the world's richest Oligocene epoch fossil beds, dating 23 to 35 million years old. The evolution of mammal species such as the horse, sheep, rhinoceros and pig can be studied in the Badlands formations.

**Operating Hours**

The park is open 24 hours a day, seven days per week. Entrance fees are collected year round.

### Big Cypress National Preserve

| | |
|---|---|
| PARKS Program: | "SWAMP: Swamp Water and Me Program" |
| Program Partner(s): | 6th-grade teachers with Collier County, Florida Public Schools |
| Subjects of Study: | Environmental science; vegetation and animal identification |
| Address: | Big Cypress National Preserve<br>HCR 61, Box 110<br>Ochopee, FL 33141 |
| Phone: | 941-695-2000 |
| Fax: | 941-695-3007 |
| Email: | sandy_snell-dobert@nps.gov |
| URL: | www.nps.gov/bicy/ |

**Appendix B**

**National Park Service Description**
Big Cypress National Preserve was set aside in 1974 to ensure the preservation, conservation, and protection of the natural scenic, floral and faunal, and recreational values of the Big Cypress Watershed. The importance of this watershed to the Everglades National Park was a major consideration for its establishment. The name Big Cypress refers to the large size of this area. Vast expanses of cypress strands span this unique landscape.

**Operating Hours**
Daily except December 25, 8:30 a.m. to 4:30 p.m.

## Big Thicket National Preserve

| | |
|---|---|
| PARKS Program: | "Ecosystem Analysis Program" |
| Program Partner(s): | Middle and high school teachers |
| Subjects of Study: | Environmental science: atmosphere/climate, hydrology, soils, and land cover/biology |
| Address: | Big Thicket National Preserve<br>3785 Milam<br>Beaumont, TX 77701 |
| Phone: | 409-839-2689 |
| Fax: | 409-839-2599 |
| Email: | BITH_Administration@nps.gov |
| URL: | www.nps.gov/bith/ |

**National Park Service Description**
The Preserve consists of nine land units and six water corridors encompassing more than 97,000 acres. Big Thicket was the first Preserve in the National Park System established October 11, 1974, and protects an area of rich biological diversity. A convergence of ecosystems occurred here during the last Ice Age. It brought together, in one geographical location, the eastern hardwood forests, the Gulf coastal plains, and the midwest prairies. In 1981, the Preserve was designated an International Biosphere Reserve by the United Nations Education, Scientific and Cultural Organization (UNESCO) Man and the Biosphere Program.

**Operating Hours**
Preserve Headquarters is open 8:00 a.m. to 4:30 p.m., Monday through Friday; closed on all government holidays. Staley Cabin Information Station is open 9:00 a.m. to 5:00 p.m., daily. Closed on December 25 and January 1.

## Appendix B

### Canyon de Chelly National Monument

| | |
|---|---|
| PARKS Program: | "Learning from the Canyon" |
| Program Partner(s): | 1st–3rd-grade teachers at Chinle Primary School, Chinle, Arizona |
| Subjects of Study: | Erosion, native (Navajo) plants, park preservation, seasonal changes |
| Address: | Canyon de Chelly National Monument<br>PO Box 588<br>Chinle, AZ 86503 |
| Phone: | 520-674-5500 |
| Fax: | 520-674-5507 |
| Email: | CACH_Superintendent@nps.gov |
| URL: | *www.nps.gov/cach/* |

**National Park Service Description**
At the base of sheer red cliffs and in-anyon wall caves are ruins of Indian villages built between AD 350 and 1300. Canyon de Chelly National Monument offers visitors the chance to learn about Southwestern Indian history from the earliest basketmakers to the Navajo Indians who live and farm here. Authorized April 1, 1931. Boundary change: March 1,1933. Acreage: 83,840— all nonfederal.

**Operating Hours**
The Visitor Center is open daily from 8 a.m. to 5 p.m., October to April; and 8 a.m. to 6 p.m., May to September.

### Channel Islands National Park

| | |
|---|---|
| PARKS Program: | "Scientists in Training" |
| Program Partner(s): | Middle and high school teachers in Ventura, California Public Schools |
| Subjects of Study: | Resource management, island ecology |
| Address: | Channel Islands National Park<br>1901 Spinnaker Drive<br>Ventura, CA 93001 |
| Phone: | 805-658-5700 |
| Fax: | 805-658-5799 |
| Email: | chis_interpretation@nps.gov |
| URL: | *www.nps.gov/chis/* |

**National Park Service Description**
Comprised of five in a chain of eight southern California islands near Los Angeles, Channel Islands National Park is home to a wide variety of nationally and

internationally significant natural and cultural resources. Over 2,000 species of plants and animals can be found within the park. However, only four mammals are endemic to the islands. One hundred and forty-five of these species are unique to the islands and found nowhere else in the world. Marine life ranges from microscopic plankton to the endangered blue whale. Archeological and cultural resources span a period of more than 10,000 years. The park consists of 249,354 acres, half of which are under the ocean. Even though the islands seem close to the densely populated, southern California coast, their isolation has left them relatively undeveloped, making them an exciting place for visitors to explore.

**Operating Hours**
The park is open all year. The Robert J. Lagomarsino Visitor Cnter hours are: Labor Day through Memorial Day, 8:30 to 4:30 weekdays, 8:00 to 5:00 weekends; Memorial Day through Labor Day, 8:00 to 5:00 weekdays, 8:00 to 5:30 weekends. The Visitor Center is closed Thanksgiving Day and December 25.

## Cuyahoga Valley National Recreation Area

| | |
|---|---|
| PARKS Program: | "Watershed Connections"(Workshop Host) |
| Program Partner(s): | 8th-grade teachers registered with the Cuyahoga Valley Environmental Education Center (CVEEC) |
| Subjects of Study: | Environmental science |
| Address: | Cuyahoga Valley National Recreation Area<br>15610 Vaughn Road<br>Brecksville, OH 44141-3018 |
| Phone: | 216-524-1497 |
| Fax: | 440-546-5905 |
| Email: | cuva _canal_visitor_center@nps.gov |
| URL: | www.nps.gov/cuva/ |

**National Park Service Description**
Cuyahoga Valley National Recreation Area, located between Cleveland and Akron, features a wide variety of natural, cultural, and historical resources. The National Park Service manages the park in cooperation with others who own property within its boundaries, including Cleveland Metroparks and Metro Parks, Serving Summit County, both of which administer several units within CVNP. Together they protect the natural landscape, preserve remnants of the area's human history, and provide a place where you can relax, play and learn new things in a beautiful outdoor setting.

**Operating Hours**
Open daily, dawn to dusk; All visitor centers are closed Thanksgiving Day, December 25, and January 1.

## Appendix B

## Delaware Water Gap National Recreation Area

| | |
|---|---|
| PARKS Program: | Environmental Science Education through PARKS Leadership Institutes (Workshop Host) |
| Program Partner(s): | Teachers throughout New York, New Jersey, and Pennsylvania; Global Learning and Observations to Benefit the Environment (GLOBE) |
| Subjects of Study: | Environmental science |
| Address: | Delaware Water Gap National Recreation Area<br>1 River Road<br>Bushkill, PA 18324 |
| Phone: | 570-588-2435 |
| Fax: | 570-588-2780 |
| Email: | dewa_interpretation@nps.gov |
| URL: | *www.nps.gov/dewa/* |

**National Park Service Description**
Delaware Water Gap National Recreation Area was established on September 1, 1965, for public outdoor recreation use and for the preservation of scenic, scientific and recreation resources. The almost 70,000 acre National Park Service unit is located in New Jersey and Pennsylvania along approximately 40 miles of the Delaware River. The area is rich in both cultural and natural history; the ridges and river valleys contain streams, waterfalls, geologic features, a diversity of plants and wildlife, and traces of past occupants and cultures. The area also provides a wide range of recreational activities.

**Operating Hours**
Most roadways and the Delaware River are open 24 hours a day, year-round, unless closed due to heavy snowfall, ice, or other hazardous conditions. River accesses and most facilities open dawn to dusk, 365 days of the year. Some facilities are closed in winter.

## Dinosaur National Monument

| | |
|---|---|
| PARKS Program: | Uinta Basin Ecosystem Program |
| Program Partner(s): | Colorado State University; local educators |
| Subjects of Study: | Environmental education; ecosystems |
| Address: | Dinosaur National Monument<br>4545 E. Highway 40<br>Dinosaur, CO 81610-9724 |
| Phone: | 970-374-3000 |
| Fax: | 970-374-3003 |
| Email: | DINO_Superintendent@nps.gov |
| URL: | *www.nps.gov/dino/* |

**National Park Service Description**
John Wesley Powell named this area Echo Park in 1869 during his first scientific expedition into the Colorado Plateau. It is here that the Yampa River, the last free-flowing river in the Colorado River System, joins the Green River. This is home and critical habitat for the endangered peregrine falcon, bald eagle, Colorado pikeminnow, and razorback sucker. Visitors can see dinosaur bones at nearby Dinosaur Quarry and explore the rugged mountain and canyon country in this park's 210,000 acres. Indian rock art in Echo Park testifies to the allure these canyons and rivers had for prehistoric people.

**Operating Hours**
The Headquarters Visitor Center is open 8 a.m. to 4:30 p.m. weekdays and is closed on federal holidays during fall, winter, and spring months. Open 8 a.m. to 6:00 p.m. on weekends during summer. Trails, auto tours, campgrounds, and backcountry areas are open all year unless closed by weather conditions.

## Edison National Historic Site

| | |
|---|---|
| PARKS Program: | "The Talking Wonder" (provisional title) |
| Program Partner(s): | Educators in New York and New Jersey |
| Subjects of Study: | Technology, scientific process |
| Address: | Edison National Historic Site<br>Main Street and Lakeside Avenue<br>West Orange, NJ 07052 |
| Phone: | 973-736-0550 |
| Fax: | 973-736-8496 |
| Email: | EDIS_Superintendent@nps.gov |
| URL: | www.nps.gov/edis/ |

**National Park Service Description**
For more than 40 years, the laboratory created by Thomas Alva Edison in West Orange, New Jersey, had enormous impact on the lives of millions of people worldwide. Out of the West Orange laboratories came the motion picture camera, vastly improved phonographs, sound recordings, silent and sound movies, and the nickel-iron alkaline electric storage battery. Edison National Historic Site provides a unique opportunity to interpret and experience important aspects of America's industrial, social, and economic past, and to learn from the legacy of the world's best-known inventor.

**Operating Hours**
The visitor center contains exhibits about Edison's work with the phonograph, motion pictures, electricity, cement, and the storage battery. The visitor center is open from 12:00 to 5:00 pm on Wednesday, Thursday, and Friday. Saturday and Sunday the visitor center is open from 9:00 am to 5:00 pm.

## Appendix B

## Fort Clatsop National Memorial

| | |
|---|---|
| PARKS Program: | "Natural History of the Lewis and Clark Expedition: Scientific Discovery and the Impacts of Change" |
| Program Partner(s): | 3rd–8th-grade teachers |
| Subjects of Study: | Environmental impact, history |
| Address: | Fort Clatsop National Memorial<br>92343 Fort Clatsop Road<br>Astoria, OR 97103-9197 |
| Phone: | 503-861-2471 |
| Fax: | 503-861-2585 |
| Email: | FOCL_Superintendent@nps.gov |
| URL: | *www.nps.gov/focl/* |

**National Park Service Description**

This site celebrates the 1805–06 winter encampment of the 33-member Lewis and Clark Expedition. A 1955 community-built replica of the explorers' 50' x 50' Fort Clatsop is the focus of this 125-acre park. The fort, historic canoe landing, and spring are nestled in the coastal forests and wetlands of the Coast Range as it merges with the Columbia River Estuary. The Salt Works unit commemorates the expedition's salt-making activities. Salt obtained from seawater was essential to the explorers' winter at Fort Clatsop and their journey back to the United States in 1806.

**Operating Hours**

Daily, summer: 8:00 a.m. to 6:00 p.m.; daily, winter: 8:00 a.m. to 5:00 p.m.; closed December 25.

## Fort Sumter National Monument

| | |
|---|---|
| PARKS Program: | "The Science of Rice Culture at Snee Farm" |
| Program Partner(s): | Elementary teachers |
| Subjects of Study: | Water quality studies, soils, plants, tides, historical agricultural practices, and the role of West African enslaved people in early American agriculture |
| Address: | Fort Sumter National Monument<br>1214 Middle Street<br>Sullivan's Island, SC 29482 |
| Phone: | 843-883-3123 |
| Fax: | 843-883-3910 |
| Email: | FOSU_Ranger_Activities@nps.gov |
| URL: | *www.nps.gov/fosu/* |

**Appendix B**

**National Park Service Description**
The first engagement of the Civil War took place at Fort Sumter on April 12 and 13, 1861. After 34 hours of fighting, the Union surrendered the fort to the Confederates. From 1863 to 1865, the Confederates at Fort Sumter withstood a 22-month siege by Union forces. During this time, most of the fort was reduced to brick rubble. Fort Sumter became a national monument in 1948. From before the American Revolution, to the close of World War II, the parks of Fort Sumter Group encompass American history in its entirety.

**Operating Hours**
Daily, except January 1 and December 25. Hours: 10:00 a.m. to 5:30 p.m., April through Labor Day, and 10:00 a.m. to 4:00 p.m. March and September through November. Hours vary otherwise and can be determined by contacting the park.

## Gateway National Recreation Area

| | |
|---|---|
| PARKS Program: | "Coastal Discoveries" |
| Program Partner(s): | 5th-grade teachers in Staten Island, New York Public Schools; Staten Island Science Teachers Association |
| Subjects of Study: | Coastal ecosystems, wildlife adaptations |
| Address: | Public Affairs Office<br>Floyd Bennett Field, Building 69<br>Brooklyn, NY 11234 |
| Phone: | 718-338-3338 |
| Fax: | 718-338-6284 |
| Email: | carole_silano@nps.gov |
| URL: | *www.nps.gov/gate/* |

**National Park Service Description**
Gateway NRA is a 26,000-acre recreation area located in the heart of the New York Metropolitan area. The park extends through three New York City boroughs and into northern New Jersey. Contact a Gateway visitor center for special programs for school groups and summer activity programs.

**Operating Hours**
Great Kills Park Ranger Station: Open all year 8:30 to 5 p.m. Located at Hylan Boulevard Staten Island, New York, 718-987-6790.

## George Washington Birthplace National Monument

| | |
|---|---|
| PARKS Program: | "How Math and Science Changed George Washington's Life" |
| Program Partner(s): | 6th & 7th-grade teachers in Richmond, Virginia County Schools |

## Appendix B

Subjects of Study: Math, environmental science, history, and geography
Address: George Washington Birthplace NM
1732 Popes Creek Road
Washington's Birthplace, VA 22443-5115
Phone: 804-224-1732
Fax: 804-224-2142
Email: GEWA_Superintendent@nps.gov
URL: *www.nps.gov/gewa/*

**National Park Service Description**
George Washington Birthplace National Monument has some of the most interesting and scenic treasures in Virginia. The 550-acre park lies within the Chesapeake Bay eco-system and includes Potomac River beach, upland forest, open fields, and marshlands. The park contains unique plant and animal species. It also contains a variety of cultural resources from the Middle Woodland prehistoric period, Henry Brooks' house site (1651-1725), John Washington and descendents' house site and adjacent family burial ground (1656-1743), and Augustine Washington's house site (George Washington's birthplace, 1718-1779). These resources are under study by archeologists from the National Park Service and other institutions.

**Operating Hours**
9:00 a.m. to 5:00 p.m. daily. Closed December 25 and January 1. Visitation highest in Summer. School group visit highest in April and May.

## Glen Canyon National Recreation Area

PARKS Program: "Field Science Partnership"
Program Partner(s): Page, Arizona Unified School District
Subjects of Study: Scientific method, natural resource management/monitoring
Address: Glen Canyon National Recreation Area
PO Box 1507
Page, AZ 86040-1507
Phone: 520-608-6200
Fax: 520-608-6283
Email: GLCA_CHVC@nps.gov
URL: *www.nps.gov/glca/*

**National Park Service Description**
Glen Canyon National Recreation Area (NRA) offers unparalleled opportunities for water-based and backcountry recreation. The recreation area stretches for hundreds of miles from Lees Ferry in Arizona to the Orange Cliffs of southern Utah, encompassing scenic vistas, geologic wonders, and a panorama of human history. Additionally, the controversy surrounding the construction of Glen Canyon Dam

and the creation of Lake Powell contributed to the birth of the modern-day environmental movement.

**Operating Hours**
Carl Hayden Visitor Center, Page, Arizona: daily, Memorial Day to Labor Day, 7 a.m. to 7 p.m.; rest of year, daily, 8 a.m. to 5 p.m.; closed Thanksgiving Day, December 25 and January 1.

Bullfrog Visitor Center, Bullfrog, Utah: intermittently in March, daily April to October, 8 a.m. to 5 p.m.; closed November to February.

Navajo Bridge Interpretive Center, near Lees Ferry: daily mid-April to October, 9 a.m. to 5 p.m.; weekends only, early April and November, 10 a.m. to 4 p.m.

## Golden Gate National Recreation Area

| | |
|---|---|
| PARKS Program: | "Here's the Dirt: Science Education at the Native Plant Nursery" |
| Program Partner(s): | Middle school teachers; AmeriCorps Volunteers |
| Subjects of Study: | Resource management, ESL |
| Address: | Golden Gate National Recreation Area<br>Fort Mason, Building 201<br>San Francisco, CA 94123-0022 |
| Phone: | 415-556-0560 |
| Fax: | 415-561-4750 |
| Email: | George_Su@nps.gov |
| URL: | www.nps.gov/goga/ |

**National Park Service Description**
The Golden Gate National Recreation Area (GGNRA) is the largest urban national park in the world. The total park area is 74,000 acres of land and water; approximately 28 miles of coastline lie within its boundaries. Golden Gate NRA has more information on specific sites such as Alcatraz, Marin Headlands, Fort Funston, Fort Mason, as well as Muir Woods National Monument, Fort Point National Historic Site, and the Presidio of San Francisco—each with its own unique natural, cultural, and military histories.

**Operating Hours**
Access to the recreation area is possible daily and year-round. Hours for individual facilities vary, but the most common hours of operation are 10:00 a.m. to 5:00 p.m. All visitor centers are open daily except for Thanksgiving Day, December 25, and January 1.

Appendix B

## Great Smoky Mountains National Park

PARKS Program: "Discovering Diversity: Integrated Science in the Smokies"
Program Partner(s): K–7 teachers in local schools, Pi Beta Phi, and the Great Smoky Mountains Institute at Tremont
Subjects of Study: Biodiversity, species identification, ecosystem monitoring
Address: Great Smoky Mountains National Park
107 Park Headquarters Road
Gatlinburg, TN 37738
Phone: 865-436-1200
Fax: 865-436-1220
Email: grsm_smokies_information@nps.gov
URL: *www.nps.gov/grsm/*

**National Park Service Description**
The Great Smoky Mountains National Park, in the states of North Carolina and Tennessee, encompasses 800 square miles of which 95 percent are forested. World-renowned for the diversity of its plant and animal resources, the beauty of its ancient mountains, the quality of its remnants of Southern Appalachian mountain culture, and the depth and integrity of the wilderness sanctuary within its boundaries, it is one of the largest protected areas in the East.

**Operating Hours**
The park is open year-round. Visitor centers at Sugarlands and Oconaluftee are open all year, except December 25 Day. Cades Cove Visitor Center has limited winter hours.

## Hagerman Fossil Beds National Monument

PARKS Program: "The Principles of Science—A Virtual Classroom"
Program Partner(s): Idaho science educators
Subjects of Study: Paleontology, biology, geology, botany, archaeology, resource management
Address: Hagerman Fossil Beds National Monument
PO Box 570, 221 North State Street
Hagerman, ID 83332
Phone: 208-837-4793
Fax: 208-837-4857
Email: HAFO_Superintendent@nps.gov
URL: *www.nps.gov/hafo/*

**National Park Service Description**
Hagerman Fossil Beds NM contains the largest concentration of Hagerman Horse fossils in North America. The Monument is internationally significant because it

protects the world's richest known fossil deposits from the late Pliocene epoch, 3.5 million years ago. These plants and animals represent the last glimpse of time that existed before the Ice Age, and the earliest appearances of modern flora and fauna.

**Operating Hours**
Summer: 9:00 a.m. to 5:00 p.m. (from Memorial Day weekend to Labor Day weekend); 10:00 a.m. to 4:00 p.m. Thursday, Friday, and Saturday (the remainder of the year).

## Indiana Dunes National Lakeshore

| | |
|---|---|
| PARKS Program: | "High School Program Development Project" |
| Program Partner(s): | High school teachers enrolled in the residential high school program at Indiana Dunes Environmental Learning Center |
| Subjects of Study: | Water quality, ecosystems |
| Address: | Indiana Dunes National Lakeshore 1100 N. Mineral Springs Road Porter, IN 46304 |
| Phone: | 219-926-7561 x225 |
| Fax: | 219-926-7561 |
| Email: | INDU_Interpretation@nps.gov |
| URL: | www.nps.gov/indu/ |

**National Park Service Description**
Indiana Dunes National Lakeshore is located approximately 50 miles southeast of Chicago, Illinois. The national lakeshore runs for nearly 25 miles along southern Lake Michigan. The park contains approximately 15,000 acres, 2,182 of which are located in Indiana Dunes State Park and managed by the Indiana Department of Natural Resources. Miles of beaches, sand dunes, bog, wetlands, woodland forests, an 1830s French Canadian homestead, and a working 1900-era farm combine to make the national lakeshore a unique setting for studying humans and their impact on the environment. Indiana Dunes is ranked seventh among national parks in native plant diversity.

**Operating Hours**
Park Headquarters: 8:00 a.m. to 4:30 p.m., Monday through Friday.

## Joshua Tree National Park

| | |
|---|---|
| PARKS Program: | "Mobile Laboratory" |
| Program Partner(s): | K–12 teachers in 1800-square-mile area |
| Subjects of Study: | Geology, mapping, animals, plants, air quality, and archaeology |

## Appendix B

| | |
|---|---|
| Address: | Joshua Tree National Park<br>74485 National Park Drive<br>Twentynine Palms, CA 92277-3597 |
| Phone: | 760-367-5500 |
| Fax: | 760-367-6392 |
| Email: | JOTR_Info@nps.gov |
| URL: | www.nps.gov/jotr/ |

**National Park Service Description**

Two deserts, two large ecosystems whose characteristics are determined primarily by elevation, come together at Joshua Tree National Park. Below 3,000 feet, the Colorado Desert encompasses the eastern part of the park and features natural gardens of creosote bush, ocotillo, and cholla cactus. The higher, moister, and slightly cooler Mojave Desert is the special habitat of the Joshua tree. In addition to Joshua tree forests, the western part of the park also includes some of the most interesting geologic displays found in California's deserts. Five fan palm oases also dot the park, indicating those few areas where water occurs naturally and wildlife abounds.

**Operating Hours**

The park is always open. Visitor centers are open daily from 8 a.m. to 4 p.m.

## Kenai Fjords National Park

| | |
|---|---|
| PARKS Program: | "Sea Life in the Fjords" |
| Program Partner(s): | Alaska Sea Life Center; K–12 educators |
| Subjects of Study: | Sea life |
| Address: | National Park Service<br>PO Box 1727<br>Seward, AK 99664 |
| Phone: | 907-224-3175 |
| Fax: | 907-224-2144 |
| Email: | KEFJ_Superintendent@nps.gov |
| URL: | www.nps.gov/kefj/ |

**National Park Service Description**

Located on the southeastern Kenai Peninsula, the national park is a pristine and rugged land supporting many unaltered natural environments and ecosystems. The park's wildlife includes mountain goats, moose, bears, wolverines, marmots, and other land mammals who have established themselves on a thin life zone between marine waters and the icefield's frozen edges. Bald eagles nest in the tops of spruce and hemlock trees. Thousands of seabirds, including puffins, kittiwakes, and murres seasonally inhabit the steep cliffs and rocky shores. Kayakers, fishermen, and visitors on tour boats share the park's waters with stellar sea lions, harbor seals, Dall porpoises, sea otters, humpback, killer, and minke whales.

**Operating Hours**
The visitor center in Seward is open Monday through Friday year round; Saturdays and Sundays from Memorial Day through Labor Day, with extended hours. Rangers provide information daily during the summer months at the ranger station at Exit Glacier.

## Kenilworth Park and Aquatic Gardens (National Capital Parks-East)

PARKS Program: "Parks Learning Exploration about Science"
Program Partner(s): K–8 teachers from three local schools
Subjects of Study: Wetlands, amphibians, art, and creative writing
Address: Kenilworth Park and Aquatic Gardens
National Capital Parks—East
1900 Anacostia Drive, S.E.
Washington, DC 20020
Phone: 202-690-5185
Email: nace_superintendent@nps.gov
URL: *www.nps.gov/kepa/*

**National Park Service Description**
Kenilworth Park and Aquatic Gardens constitutes some 700 acres and is part of Anacostia Park. The origins of Kenilworth Park and Aquatic Gardens lie not only in the 1791 L'Enfant Plan for the District of Columbia, but also the McMillan Plan of 1901 which specifically recommended extension of public parkland along both sides of the Anacostia River. The Kenilworth Aquatic Gardens is the only National Park Service site devoted to the propagation and display of aquatic plants. The Gardens were begun as the hobby of a Civil War veteran and operated for 56 years as a commercial water garden. In 1938, the Gardens were purchased from by the federal government and became part of the National Park system.

**Operating Hours**
Open every day of the year except Thanksgiving Day, December 25, and January 1.

## Lake Meredith National Recreation Area

PARKS Program: "Public Landcorps Cooperative Education Program"
Program Partner(s): Public Landcorps and local schools in Fritch, Texas
Subjects of Study: Ecology, communities, water and soil quality testing
Address: Lake Meredith National Recreation Area
419 E. Broadway
Fritch, TX 79036
Phone: 806-857-3151
Fax: 806-857-2319

## Appendix B

Email: LAMR_Interpretation@nps.gov
URL: www.nps.gov/lamr/

**National Park Service Description**
Lake Meredith lies on the dry and windswept High Plains of the Texas Panhandle in a region known as Llano Estacado, or Staked Plain. Lake Meredith was created by Sanford Dam on the Canadian River and now fills many breaks whose walls are crowned with white limestone caprock, scenic buttes, pinnacles, and red-brown, wind-eroded coves. Above lies the mesquite, prickly pear, yucca, and grasses of arid plains. And up the sheltered creek beds stand cottonwoods, soapberry, and sandbar willows. The National Park Service administers the recreation area under a cooperative agreement with the Bureau of Reclamation.

**Operating Hours**
Administrative Offices: 8:00 a.m. to 4:30 p.m. Park open 24 hours.

## Little River Canyon National Preserve

PARKS Program: "Thinking Big with Little River Canyon"
Program Partner(s): K–5 teachers at Donahoo Technology Center
Subjects of Study: Weather, species interdependence, human impact on Earth, hydrologic cycle, climate, water quality protection, ecosystems
Address: Little River Canyon National Preserve
2141 Gault Avenue North
Fort Payne, AL 35967
Phone: 256-845-9605
Fax: 256-997-9129
Email: LIRI_Superintendent@nps.gov
URL: www.nps.gov/liri/

**National Park Service Description**
Little River flows for most of its length atop Lookout Mountain in northeast Alabama. The river and canyon systems are spectacular Appalachian Plateau landscapes any season of the year. Forested uplands, waterfalls, canyon rims and bluffs, stream riffles and pools, boulders, and sandstone cliffs offer settings for a variety of recreational activities. Natural resources and cultural heritage come together to tell the story of the Preserve, a biologically diverse area with a number of rare plants and animals such as the green pitcher plant, and endangered fish called blue shiner and the green salamander.

**Operating Hours**
The park is open year-round. The Canyon Mouth Park Unit is a day-use area.

## Lowell National Historical Park

| | |
|---|---|
| PARKS Program: | "Tsongas Industrial History Center Curriculum Redesign" (Workshop Host) |
| Program Partner(s): | University of Massachusetts, Lowell |
| Subjects of Study: | Social studies, history, technology, and science |
| Address: | Lowell National Historical Park<br>67 Kirk Street<br>Lowell, MA 01852 |
| Phone: | 978-970-5000 |
| Fax: | 978-275-1762 |
| Email: | LOWE_Superintendent@nps.gov |
| URL: | www.nps.gov/lowe/ |

**National Park Service Description**

The history of America's Industrial Revolution is commemorated in Lowell, Massachusetts. The Boott Cotton Mills Museum with its operating weave room of 88 power looms, "mill girl" boardinghouses, the Suffolk Mill Turbine Exhibit, and guided tours tell the story of the transition from farm to factory, chronicle immigrant and labor history, and trace industrial technology. The park includes textile mills, worker housing, 5.6 miles of canals, and 19th-century commercial buildings.

**Operating Hours**

The Visitor Center is open year-round 9:00 a.m. until 5:00 p.m. Visitor Center hours are extended during the summer season. The Boott Cotton Mills Museum and Working People Exhibit hours vary by season. Call ahead for specific exhibit hours. Park is closed Thanksgiving Day, December 25, and January 1.

## Point Reyes National Seashore

| | |
|---|---|
| PARKS Program: | "Creating Coastal Stewardship Through Science" |
| Program Partner(s): | Middle school teachers in Marin County, California Public Schools |
| Subjects of Study: | Adaptations, migration, oceanography, ocean habitats, geology, land habitats, resident species |
| Address: | Point Reyes National Seashore<br>Point Reyes, CA 94956 |
| Phone: | 415-464-5100 |
| Fax: | 415-663-8132 |
| Email: | PORE_Webmaster@nps.gov |
| URL: | www.nps.gov/pore/ |

# Appendix B

**National Park Service Description**
Point Reyes National Seashore contains unique elements of biological and historical interest in a spectacularly scenic panorama of thunderous ocean breakers, open grasslands, bushy hillsides, and forested ridges. Native land mammals number about 37 species, and marine mammals augment this total by another dozen species. The biological diversity stems from a favorable location in the middle of California and the natural occurrence of many distinct habitats. Nearly 20 percent of the State's flowering plant species are represented on the peninsula and over 45 percent of the bird species in North America have been sighted. The Point Reyes National Seashore was established by President John F. Kennedy on September 13, 1962.

**Operating Hours**
The park is open daily (with overnight camping available by permit only) from sunrise to sunset throughout the year. Visitor center hours vary; contact the park for more information. All visitor centers are closed December 25.

## Redwood National and State Parks

| | |
|---|---|
| PARKS Program: | "Resource Stewardship Traveling Trunk" |
| Program Partner(s): | Middle school teachers in local schools |
| Subjects of Study: | Forest ecology, wildlife, recreation and watershed management |
| Address: | Redwood National and State Parks<br>1111 Second Street<br>Crescent City, CA 95531 |
| Phone: | 707-464-6101 |
| Fax: | 707-464-1812 |
| Email: | REDW_Information@nps.gov |
| URL: | www.nps.gov/redw/ |

**National Park Service Description**
Redwood National and State Parks are home to some of the world's tallest trees: old-growth coast redwoods. They can live to be 2,000 years old and grow to over 300 feet tall. The parks' mosaic of habitats include prairie/oak woodlands, mighty rivers and streams, and 37 miles of pristine Pacific coastline. Cultural landscapes reflect American Indian history. The more recent logging history has led to much restoration of these parks. Three California state parks and the National Park Service unit represent a cooperative management effort of the National Park Service and California Department of Parks and Recreation. Together these parks are a World Heritage Site and International Biosphere Reserve, protecting resources cherished by citizens of many nations.

**Operating Hours**
Redwood is accessible year-round. Heavy rains can lead to road closures; check at a visitor center to find out the estimated open date.

# Rocky Mountain National Park

| | |
|---|---|
| PARKS Program: | "Heart of the Rockies Education Program" |
| Program Partner(s): | K–9 teachers; North American Association for Environmental Education; Colorado State University; Front Range Community College; Poudre School District |
| Subjects of Study: | Environmental science; interdisciplinary studies |
| Address: | Rocky Mountain National Park<br>1000 Highway 36<br>Estes Park, CO 80517-8397 |
| Phone: | 970-586-1206 |
| Fax: | 970-586-1256 |
| Email: | ROMO_Information@nps.gov |
| URL: | www.nps.gov/romo/ |

**National Park Service Description**

The park's rich scenery typifies the massive grandeur of the Rocky Mountains. Trail Ridge Road crosses the Continental Divide and looks out over peaks that tower more than 14,000 feet high. Wildlife and wildflowers call these 415.2 square miles (265,727 acres) of Colorado's front range, home. Also preserved within the park boundaries are some of Colorado's more pristine forests. Great stands of ponderosa pine, Douglas fir, lodgepole pine, aspen, subalpine fir and spruce adorn the mountains below treeline. The forests are interspersed with mountain meadows that fill with colorful wildflowers during the brief high country summer. The road is closed over the winter.

**Operating Hours**

Open 24 hours a day, 12 months of the year. Pets are allowed in campgrounds, picnic areas, and along roadsides, but not on any park trails. Pets cannot be left in cars unattended.

# Santa Monica Mountains National Recreation Area

| | |
|---|---|
| ARKS Program: | "Nature's Laboratories in PARKS" |
| Program Partner(s): | K–12 teachers |
| Subjects of Study: | Native populations; air, soil, and water; wildland fire ecology |
| Address: | Santa Monica Mountains National Recreation Area<br>401 West Hillcrest Drive<br>Thousand Oaks, CA 91360 |
| Phone: | 805-370-2300 |
| Fax: | 805-370-1850 |
| Email: | SAMO_interpretation@nps.gov |
| URL: | www.nps.gov/samo/ |

## Appendix B

**National Park Service Description**
Santa Monica Mountains rise above Los Angeles, widen to meet the curve of Santa Monica Bay, and reach their highest peaks facing the ocean, forming a beautiful and multi-faceted landscape. Santa Monica Mountains National Recreation Area is a cooperative effort that joins federal, state, and local park agencies with private preserves and landowners to protect the natural and cultural resources of this transverse mountain range and seashore. Located in a Mediterranean ecosystem, the Santa Monica Mountains contain a wide variety of plants and wildlife. The mountains also have an interesting and diverse cultural history, which begins with the Chumash and Gabrielino/Tongva peoples and continues today in "L.A.'s backyard."

**Operating Hours**
The National Park Service Visitor Center is open daily from 9 a.m. to 5 p.m. It is closed on Thanksgiving Day, December 25, and January 1. The Visitor Center contains a bookstore, which is open during the same hours.

## Shenandoah National Park

| | |
|---|---|
| PARKS Program: | "Integrated Science Education Program" |
| Program Partner(s): | McGaheysville Elementary School (Virginia) |
| Subjects of Study: | Habitats, environmental science |
| Address: | Shenandoah National Park<br>3655 U.S. Highway 211 East<br>Luray, VA 22835-9036 |
| Phone: | 540-999-3500 |
| Fax: | 540-999-3601 |
| Email: | SHEN_Superintendent@nps.gov |
| URL: | www.nps.gov/shen/ |

**National Park Service Description**
Shenandoah National Park holds more than 500 miles of trails, including 101 miles of the Appalachian Trail. Trails may follow a ridge crest, or they may lead to high places with panoramic views or to waterfalls in deep canyons. Skyline Drive, a 105-mile road that winds along the crest of the mountains through the length of the park, provides vistas of the spectacular landscape to east and west. Many animals, including deer, black bears, and wild turkeys, flourish among the rich growth of an oak-hickory forest. In season, bushes and wildflowers bloom along the Drive and trails and fill the open spaces. Apple trees, stone foundations, and cemeteries are reminders of the families who once called this place home.

**Operating Hours**
The park itself is always open, but some portions of the Skyline Drive, the only road through Shenandoah National Park, are closed from dusk to early morning during

hunting season. This road also closes in inclement weather for safety reasons. Visitor facilities and services begin operating between early April and Memorial Day and close down by late November.

## Sleeping Bear Dunes National Lakeshore

PARKS Program: "Ecological Restorations of Disturbed Lands"
Program Partner(s): Glen Lake High School (Michigan)
Subjects of Study: Environmental impact and restoration, native plants, soils
Address: Sleeping Bear Dunes National Lakeshore
9922 Front Street
Empire, MI 49630-9797
Phone: 231-326-5134
Fax: 231-326-5382
Email: SLBE_Interpretation@nps.gov
URL: www.nps.gov/slbe/

**National Park Service Description**

Sleeping Bear Dunes National Lakeshore encompasses a 60 km (35 mi.) stretch of Lake Michigan's eastern coastline, as well as North and South Manitou Islands. The park was established primarily for its outstanding natural features, including forests, beaches, dune formations, and ancient glacial phenomena. The Lakeshore also contains many cultural features, including lighthouse built in 1871, three former Life-Saving Service/Coast Guard Stations, and an extensive rural historic farm district.

**Operating Hours**

The park operates on a year-round basis. The Philip A. Hart Visitor Center is open seven days a week with the exception of federal holidays during the winter months. Summer hours are 9:00 a.m. to 6:00 p.m.; the remainder of year the Center is open 9:00 a.m. to 4:00 p.m. The Maritime Museum is open seasonally from 10:30 a.m. to 5:00 p.m.; the grounds are open year-round.

## Southeast Utah Group, Arches and Canyonlands National Parks

PARKS Program: "Canyon Country Outdoor Education"
Program Partner(s): Pre-K–12 teachers
Subjects of Study: Environmental science
Address: Southeast Utah Group
PO Box 907
Moab, UT 84532-0907
Phone: 435-719-2100
Fax: 435-719-2300

## Appendix B

Email: archinfo@nps.gov
canyinfo@nps.gov
URL: www.nps.gov/arch/ and www.nps.gov/cany/

**National Park Service Description**
Arches National Park preserves over two thousand natural sandstone arches, including the world-famous Delicate Arch, in addition to a variety of unique geological resources and formations. In some areas, faulting has exposed millions of years of geologic history. The extraordinary features of the park—including balanced rocks, fins and pinnacles—are highlighted by a striking environment of contrasting colors, landforms, and textures. Arches is located in southeast Utah, five miles north of Moab.

Canyonlands National Park preserves a colorful landscape of sedimentary sandstones eroded into countless canyons, mesas, and buttes by the Colorado River and its tributaries. Located in southeast Utah, the park sits in the heart of a vast basin bordered by sheer cliffs of Wingate Sandstone.

**Operating Hours**
Both parks are open year-round. The visitor centers are open daily from 8 a.m. to 4:30 p.m., with extended hours spring through fall. Visitor centers are closed December 25.

## Tumacácori National Historical Park

| | |
|---|---|
| PARKS Program: | "Culture's Effect on the Santa Cruz Valley: Past-Present-Future" |
| Program Partner(s): | Middle and high school teachers; US Fish and Wildlife Service; local landowners; University of Arizona; Arizona Dept. of Public Health; Anza Trail Coalition of Arizona |
| Subjects of Study: | Human impact on environment; environmental science; dendrochronology; ethnobotany; history of agriculture |
| Address: | Tumacácori National Historical Park<br>PO Box 67<br>Tumacacori, AZ 85640 |
| Phone: | 520-398-2341 |
| Fax: | 520-398-9271 |
| Email: | TUMA_Superintendent@nps.gov |
| URL: | www.nps.gov/tuma/ |

**National Park Service Description**
Tumacácori National Historical Park is located in the Santa Cruz River Valley in southern Arziona, which makes it a part of the Arizona Upland division of the Sonoran Desert—an ecosystem that is home to a great diversity of plant species as well as animals and

birds. The park is also home to the Juan Bautista de Anza National Historic Trail, authorized by Congress on August 15, 1990—the first such trail in the Western Region of the National Park Service. It is one of the long distance National Historic Trails in the United States. Tumacácori National Historical Park tells the story of the first Europeans who came to southern Arizona and of the native people who lived here then.

**Operating Hours**
Open 8:00 a.m. to 5:00 p.m. daily, except Thanksgiving Day and December 25.

## Tuskegee Institute National Historic Site

| | |
|---|---|
| PARKS Program: | "Recycling Rangers" |
| Program Partner(s): | 9th- and 10th-grade teachers at Booker T. Washington High School |
| Subjects of Study: | Technology, environmental science, chemistry, history |
| Address: | Tuskegee Institute National Historic Site<br>1212 Old Montgomery Road<br>Tuskegee Institute, AL 36087 |
| Phone: | 334-727-6390 |
| Fax: | 334-727-4597 |
| Email: | tuin_administration@nps.gov |
| URL: | www.nps.gov/tuin/ |

**National Park Service Description**
In 1881, Booker T. Washington became the first principal of a newly formed Normal School for Negroes in Tuskegee, Alabama. This began a lifelong quest for excellence that over saw the Growth of Tuskegee Institute. The historic campus district still retains the original buildings built by the students of the Institute, with bricks made by students in the Institute brickyard. In 1896, George Washington Carver joined the faculty and revolutionized agricultural development in the South in the early 20th century. The legacy of these two men, and the history of this great institution of higher education, have been preserved to tell the story of men and women, former slaves, who struggled to make their place in our American society. The site, located on the campus of Tuskegee University, became a part of the National Park System in 1974.

**Operating Hours**
The park is open daily, 9:00 a.m. to 5:00 p.m. It is closed Thanksgiving Day, December 25, and January 1.

## Appendix B

### Wind Cave National Park

PARKS Program: "Water in the Environment"
Program Partner(s): Local public schools
Subjects of Study: Cave hydrology, geology, ecosystems, soils
Address: Wind Cave National Park
RR 1, Box 190
Hot Springs, SD 57747-9430
Phone: 605-745-4600
Fax: 605-745-4207
Email: phyllis_cremonini@nps.gov
URL: www.nps.gov/wica/

**National Park Service Description**
One of the world's longest and most complex caves and 28,295 acres of mixed-grass prairie, ponderosa pine forest, and associated wildlife are the main features of the park. The cave is well known for its outstanding display of boxwork, an unusual cave formation composed of thin calcite fins resembling honeycombs. The park's mixed-grass prairie is one of the few remaining and is home to native wildlife, such as bison, elk, pronghorn, mule deer, coyotes, and prairie dogs.

**Operating Hours**
Park open year-round. Visitor center open daily except Thanksgiving Day and December 25. Visitor center hours vary depending on the season.

### Yellowstone National Park

PARKS Program: "Expedition Yellowstone!" (Workshop Host)
Program Partner(s): 4th–6th-grade teachers in Lamar Valley, Wyoming
Subjects of Study: Global environmental education, technology
Address: Yellowstone National Park
PO Box 168
Yellowstone National Park, WY 82190-0168
Phone: 307-344-7381
Fax: 307-344-2005
Email: yell_visitor_services@nps.gov
URL: www.nps.gov/yell/

**National Park Service Description**
Yellowstone National Park is the first and oldest national park in the world. The commanding features that initially attracted interest, and led to the preservation of Yellowstone as a national park, were geological: the geothermal phenomena (there are more geysers and hot springs here than in the rest of the world combined); the

colorful Grand Canyon of the Yellowstone River; fossil forests; and the size and elevation of Yellowstone Lake. The human history of the park is evidenced by cultural sites dating back 12,000 years. More recent history can be seen in the historic structures and sites that represent the various periods of park administration and visitor facilities development.

**Operating Hours**
Summer: Park entrances open on different dates when snow crews are able to clear the roads. The season runs from mid-April to late-October. Winter: The season runs from mid-December to mid-March. To learn the projected dates for the year visit www.nps.gov/yell/planvisit/orientation/travel/roadopen.htm.

# Appendix B